KU-318-468

ELECTROMAGNETISM FOR ENGINEERS

An Introductory Course

2nd Edition in SI/Metric Units

by

P. HAMMOND

Head of Department of Electrical Engineering
University of Southampton

PERGAMON PRESS

OXFORD · NEW YORK · TORONTO · SYDNEY
PARIS · FRANKFURT

U.K.	Pergamon Press Ltd., Headington Hill Hall, Oxford OX3 0BW, England .
U.S.A.	Pergamon Press Inc., Maxwell House, Fairview Park, Elmsford, New York 10523, U.S.A.
CANADA	Pergamon of Canada Ltd., 75 The East Mall, Toronto, Ontario, Canada
AUSTRALIA	Pergamon Press (Aust.) Pty. Ltd., 19a Boundary Street, Rushcutters Bay, N.S.W. 2011, Australia
FRANCE	Pergamon Press SARL, 24 rue des Ecoles, 75240 Paris, Cedex 05, France
FEDERAL REPUBLIC OF GERMANY	Pergamon Press GmbH, 6242 Kronberg-Taunus, Pferdstrasse 1, Federal Republic of Germany

Copyright © 1978 Pergamon Press Ltd.

*All Rights Reserved. No part of this publication may be
reproduced, stored in a retrieval system or transmitted in
any form or by any means: electronic, electrostatic,
magnetic tape, mechanical, photocopying, recording or
otherwise, without permission in writing from the
publishers*

First Edition 1964

Reprinted 1965

Second Edition 1978

British Library Cataloguing in Publication Data

Hammond, Percy
Electromagnetism for engineers.—2nd ed in SI
metric units.—(Pergamon international library).
1. Electromagnetic theory
I. Title   II. Series
530.1′41′02462   QC670   77–30279

ISBN 0–08–022103–3 (Hardcover)
ISBN 0–08–022104–1 (Flexicover)

QUEEN MARY
COLLEGE
LIBRARY

Printed in Great Britain by A. Wheaton & Co. Ltd, Exeter, Devon

Contents

APPENDICES

Preface to the Second Edition

IN recent years I have been teaching electromagnetism at the more advanced mathematical level suitable to the later years of an honours degree course in electrical engineering. When the Publishers asked me to prepare a second edition of this book I had not looked at it for a long time and wondered what I should find. Would I be forced to insert a good deal of mathematics or would I be content with a treatment which seeks to detach the physical phenomena from the notation in which they are generally presented? I was glad to find that the absence of mathematics did not worry me. In fact I found it a positive advantage, because so often an elegant mathematical formulation obscures the experimental basis of this subject. The quotation from Maxwell in the original preface still seems to me to contain wonderfully good advice. Maxwell was a gifted mathematician, but he knew that the phenomena are central.

Electromagnetism is not an esoteric subject for a few specialists. It is at the heart of electrical and electronic engineering. This book seeks to make the subject accessible to engineers and I am glad that it is being widely used. I hope that this fully revised second edition using SI units will introduce many more readers to a subject which is both profound and enormously useful.

Southampton   P. HAMMOND

Preface to the First Edition

ABOUT eight years ago I had the privilege of attending some excellent lectures to scientists on the principles of electromagnetism. I can well remember the surprise that was expressed by some of my friends, who felt it incongruous that an engineer should attend such lectures, particularly an engineer who was not a high-frequency specialist.

Much has happened in the engineering world during these last few years and I do not expect that my experience would be the same today. The tremendous expansion of electrical engineering practice has forced engineers to strengthen their grasp on the fundamental principles underlying their subject, so that when they are faced with new devices they have a firm base of understanding from which to survey new knowledge.

The pressure for a more scientific approach to electrical engineering has been strongly felt in universities and technical colleges. Teachers and students alike have become critical of the content of courses which deal mainly with some well-established branch of engineering technology. It is not surprising that amongst new courses there are many on the principles of electromagnetism. This book has been written as a teaching aid for such courses.

It is an elementary book in the sense that no knowledge of mathematics beyond simple differentiation and integration is required of the reader. The book could be used in the sixth form of schools and I hope it may be. But in writing it I have thought chiefly of first and second year students in colleges of technology and universities. I have based it on my teaching experience with

xiii

university students over a period of fourteen years at such different places as England, the U.S.A., and Malaya.

Although the mathematics is elementary it will, I hope, be found that I have not shirked the difficulties inherent in the subject. These difficulties are very real because to many students electrical terms, such for instance as flux, are vague and shadowy. I have tried to overcome this vagueness by using throughout the book the more familiar language of mechanics and by following the advice of Maxwell in his famous treatise: "We therefore avail ourselves of the labours of the mathematicians, and retranslate their results from the language of the calculus into the language of dynamics, so that our words may call up the mental image, not of some algebraical process, but of some property of moving bodies."

It is difficult in a work like this to make suitable acknowledgement to all the many people from whom I have learned. I hope silence will not be taken to imply lack of gratitude. I should like to thank especially Professor M. G. Say and Mr. R. H. Angus for their kindness in reading the manuscript and in making helpful suggestions. My thanks are also due to the Syndics of the Cambridge University Press for allowing me to use certain examination questions.

Cambridge   P. HAMMOND
August, 1963

"Great are the works of the Lord, studied
by all who have pleasure in them."
Psalm 111, verse 2

CHAPTER 1

The Science of Electromagnetism

1.1. THE WORLD OF ELECTRICITY

It is difficult to imagine what the world was like before the discovery of electricity. These words are being written under electric light. There is electricity in almost every home and every factory. Energy and information are transmitted by electricity and machinery is driven by electricity. Yet electrical engineering practice and the science that underlies it are relatively young. The first battery was made in 1799, the first transformer, generator and motor in 1831. This may seem a long time ago, but in comparison with civil and mechanical engineering, electricity is a new arrival.

The first book on electricity was published in 1600. Its author was William Gilbert, physician to Queen Elizabeth I of England. He was a capable experimenter as well as a clear thinker and he attacked the "vain imaginings of light-headed metaphysicians" who did not test their conclusions by experiments. Gilbert was the originator of "field theory", which we shall mention many times in this book. On a practical level he did much good work in developing the mariner's compass.

For about 150 years after Gilbert's work nothing much was done to advance the subject. But about 1750 Benjamin Franklin, American writer and statesman, put forward the notion that electricity is an elastic fluid which permeates ordinary matter. He showed that this "fluid" could be transferred from one body to another. Franklin was thus the originator of the idea of electric particles. The interplay between Gilbert's field theory and Franklin's particle theory has been of immense value to the development of electrical science. Franklin like Gilbert put his knowledge to immediate use. By his famous experiment of flying a kite into a thundercloud he demonstrated that thunderstorms are electrical phenomena. This led him to invent the lightning conductor without which the construction of tall buildings is very hazardous. Perhaps it should be said that Franklin's experiments were extremely dangerous and at least one professor was killed trying to do what Franklin did.

Franklin was followed by an ever-increasing band of investigators. We must mention Faraday who discovered electromagnetic induction in 1831. His idea of "lines of force" can be traced back to Gilbert's fields. 1886 is another milestone in electrical history, because in that year Hertz discovered that electricity can be sent through empty space and he thus laid the basis of radio transmission. But the most important step forward from a scientific point of view was Thomson's discovery of the electron in 1897. Whereas Franklin had thought that the electric fluid penetrated ordinary matter, it was now realized that ordinary matter consisted of electric particles, that we are living in an electrical world and are ourselves made up of electric particles and controlled by electric processes.

At this point we shall turn our back on the historical development and face our task of giving a coherent account of the principles of electromagnetism. There will be bits of history in other chapters of this book, but readers who want more than this should look at the books mentioned at the end of this chapter.

1.2. THE MEASUREMENT OF ELECTRICITY

Not everything in the world can be measured. In fact the most important things in life such as friendship and courage defy measurement. Science, however, concerns itself with measurable quantities. It is not enough to say that there is much electricity in a flash of lightning, the question is how much.*

The question of measurement looks innocent enough, but it has been the source of countless headaches to generations of electrical engineers and we shall approach it with caution. The trouble is that there are two matters mixed up in it. The first concerns the entities we wish to measure, for instance the thing called electric current in a lightning stroke, and the second concerns the size of the unit of the thing to be measured. In mechanics, for example, we find that the entity called length is a useful one and we decide to make it one of our basic entities or *dimensions*. That settles the first point. The second question then is in what units shall we measure length, and here we have a wide choice based on the length of certain pieces of metal kept in various standardizing laboratories in different parts of the world.

In mechanics it has been found by experience that three basic entities are sufficient to build a useful language for engineers These three dimensions are length, mass and time, which are given the symbols L, M, T. Other words in the language can be built up from these three. Thus area is (length)^2 and force is mass $\times$ length/(time)^2. In dealing with complicated equations it is often useful to carry out a *dimensional analysis*, which expresses both sides of the equation in terms of length, mass and time. If the two sides are not identical dimensionally, there is something wrong; it would, for instance, be absurd to equate area and force.

In electrical engineering it has been found that a language based only on length, mass and time is inadequate. At least one specifically *electrical* entity or dimension is needed, and electric current or rate of flow of electric charge is chosen for this purpose. It is generally given the symbol A. In this book we shall meet other electrical entities besides electric current and many of these have names of their own. However, all of them can be analysed into the four basic dimensions of length, mass, time and current. Thus we

---

* The electric current in a visible lightning stroke varies from about 500 A to as much as 100,000 A.

can build up a complete *electrical language* from these four dimensions.

Now at last we are in a position to ask the second part of the question: what shall be the size of our "unit" for each dimension? Shall length be measured in miles or in Angstrom units or in metres, and in what units shall we measure mass, time and current? The answers are of course arbitrary, but if we wish to co-operate with other people it is best to use an internationally agreed set of units.

In electrical engineering the agreed units are the metre, the kilogramme, the second and the ampere. The metre and the kilogramme were originally defined in terms of a certain bar of metal and a certain metal weight kept in Paris, and the second was defined in terms of the rotation of the earth. However in 1960 new standards were adopted internationally for the metre and the second in terms of the wave-length and frequency of the radiation from certain atoms. This new definition gives a very much improved accuracy (1 in 10^9). The ampere is defined in terms of the force between two wires carrying electric current.

The metre, kilogramme, second and ampere belong to the International System (SI) of units. Amongst other electrical units in this system are the volt and the ohm. Most electrical instruments are calibrated in SI units. The unit of power is the watt, and since power is a purely mechanical entity (Force × Velocity), its dimensional analysis can be carried out entirely in terms of length, mass and time, namely ML^2/T^3. The dimensional analysis of voltage can be derived from the fact that volts × amps = watts. Thus the volt has dimensions ML^2/T^3A.

There are of course other systems of units. Of these, the pound and the foot do not usually form the basis of an electrical system of units, although electrical engineers have to be familiar with their use. The horse-power (= 746 watts) is frequently used in calculations about electric motors, but it does not form the basis of a system of units.

However, there are two complete systems of electrical units which in the past were widely used and are still important. Both these systems are based on the centimetre, gramme and second (C.G.S.). They differ in their choice of the fourth, specifically electrical, quantity, one being derived from magnetic measurements and the other from electrostatic ones. The C.G.S. electromagnetic system of units (e.m.u.) has a unit of electric current which is 10 amperes. The C.G.S. electrostatic system of units (e.s.u.) has a unit of current of $1/(3 \times 10^9)$ amperes. Those who were brought up in these systems of units find them very convenient.

But both systems suffer from the fact that ordinary measuring instruments are calibrated in SI units. It would be far too costly to change all the meters and this is the overriding reason why the SI system is displacing the two C.G.S. systems. There is not much about economics in this book, but this choice of units is an example of the dominance of economic factors. It is the function of engineers to harness natural forces economically and to do for £1 what anybody can do for £10; an engineer who attempts to ignore economics will be unsuccessful. Since many books and articles use C.G.S. units, it is sometimes necessary to convert one system to another. A conversion table is given at the end of this book.

1.3. ENGINEERING APPLICATIONS OF ELECTRICITY

Because matter itself consists of electric particles a knowledge of electricity and its laws is essential in all branches of science and technology. In a sense every physicist, every industrial chemist and every metallurgist has to be an electrical engineer.

But, if we restrict ourselves to those activities commonly described as electrical engineering, we find there three types of application of electricity to human affairs which have been responsible for creating modern conditions of living, and which enable the world to support its enormous population.

The first and the most important application of electricity is to the conversion and transmission of energy. The kinetic and potential energy of water, the thermal energy of coal and oil and the nuclear energy of uranium and similar materials can be converted into electrical energy. This energy can be transmitted cheaply and conveniently to factories and homes, there to be reconverted into heat or mechanical work. Electrical energy makes possible for the ordinary citizen a standard of living which would in olden days have been possible only for those who could call on the labour of a large number of servants or slaves. Hence it has become common to judge the prosperity of a country by the size of its electrical installation divided by the number of its people.

The second type of activity of electrical engineers concerns the use of electricity in transmitting information. The telephone services are part of this activity and the submarine telephone cables which now encircle the globe are a great triumph of electrical engineering. Another electrical way of transmitting information is by radiation of electromagnetic waves through space. This forms the basis of sound radio and television, also of radio communication systems and of radar.

A relatively new but immensely important third type of electrical engineering has sprung from the increasing use of electrical means of calculation and of the control of devices and processes. Electronic computers have made possible the manipulation of data and the solution of problems which were beyond the power of the most clear-minded, energetic and patient of mathematicians. Electrical control of industrial processes has cheapened and improved many manufactured articles and it has also released people from tedious jobs.

Between them the three fields of energy conversion, transmission of information and control extend into almost every activity of human society. In all these fields electricity is supreme and electrical engineers fulfil a vital role.

1.4. MATERIALS USED IN ELECTRICAL ENGINEERING

The electrical nature of matter implies that almost every type of material can be used in electrical engineering. At the time of Franklin materials used to be divided into "electrics" and "non-electrics", according as they could be electrified by friction or not. This classification is echoed in our modern division of materials into insulators and conductors. Between them insulators and conductors of electricity enable the engineer to devise electric circuits, so that the electric current is forced to flow in certain conducting paths and is prevented by insulators from leaving these paths. The important property of materials which makes this possible is that of resistivity, which is defined as the resistance to current flow of a cube of unit side of the material. A wonderful range of resistivities is available in nature. Metals like copper have a resistivity of the order of 10^{-8} ohm-metres and insulators like glass have a resistivity of the order of 10^{12} ohm-metres. Thus engineers have available the fantastic range of 1 to 10^{20}. The physical explanation is that in metals electrons form an *electron gas* and are free to move through the atomic lattice. This makes it possible to move enormous amounts of electric charge by the application of small electric forces. In insulators, on the other hand, the electrons are fixed and the flow of electric current is virtually impossible.

Between the conductors and insulators are a group of materials known as semi-conductors. They consist of various metallic oxides and of materials like germanium and silicon, and they find important applications in control devices such as rectifiers and transistors.

1.5. THE SCIENCE OF ELECTROMAGNETISM

Even in such a brief survey as the one we are carrying out in this chapter, one stands amazed at the range of electrical engineering, the profusion of materials that are available and the progress that has been made. A newcomer may well be forgiven if he says to himself: "Let me shut this book, roll up my sleeves, grasp a soldering iron and get on with the serious business of becoming an electrical engineer." Much could indeed be achieved by such a method and is achieved daily. But it is not as simple as that. A firm of builders could design useful buildings without having on its staff anyone versed in structural analysis. But no firm of electrical engineers is likely to be successful in the design of electrical apparatus unless they have somebody who knows more than how to connect wires and solder joints or turn the pages of an electrical handbook. Somebody at least must have mastered the laws of electricity. To achieve such mastery requires a mind that is prepared to take the long view to forgo the immediate pleasure of being useful and to look forward to the wider usefulness that will come later. To learn about electricity is like learning a foreign language. It would be fun to go abroad at once without having learned the language. One can get far by signs and waving one's head and arms about. But foreign travel is much more enjoyable when one has some knowledge of the grammar and structure of the foreign language, so that one can hold proper conversations.

With all this the cautious student may agree, but the question in his mind is whether the analogy is justified. Is there in fact such a thing as an electrical language? Are there not rather a very large number of dialects which are incomprehensible to anyone who has not lived in the place for ten years? Are there not light current engineers, power engineers, electronic engineers, telephone engineers, microwave engineers, transformer engineers, fractional-horsepower motor engineers and so on in an almost endless profusion? What use is it to learn a language which is not spoken by the people to whom one wishes to talk?

Now the answer to these questions is almost too good to be true: there is only one language of electricity and it applies to all electrical phenomena. There are of course technological dialects which relate to methods of manufacture and special know-how. The layout of a factory making electronic components is different from one making motors for steel-works. But the electric laws governing the behaviour of electronic equipment or of electric motors are identical, it is not a question of similarity but of complete identity. The laws of electricity apply whether the frequency is 50 Hz or 50 MHz, whether the current is measured in micro-amperes or kilo-amperes, whether the equipment weighs a fraction of a gramme or a few hundred tonnes whether it costs ten pence or a million pounds. It is this universality which makes the electrical language one of the most useful and fascinating languages to learn.

What is the electrical language like? Put simply, electricity is concerned with the mechanics of electric charges or charged particles. As in ordinary mechanics there are in electricity two branches: electrostatics and electrodynamics. For reasons that will appear later the complete subject is called electromagnetism. In the rest of this book we shall formulate the laws of electromagnetism. We shall have to resist the temptation of launching into a description of hydro-electric schemes or colour television. We shall have to do a lot of grammar to master the language. But it will be worth it.

SUMMARY

In this introductory chapter we have looked at the historical development of electrical engineering. We have seen that there has grown up a special vocabulary of electrical terms and that systems of units have been devised so that the sizes of electrical quantities can be determined. We have seen why the SI system has been generally adopted by engineers. A brief survey of the engineering applications of electricity and of the materials used has illustrated the vast range of electrical engineering. Finally we have found that the laws of electromagnetism apply to all branches of the subject and that this universality makes a study of these laws worth while.

SUGGESTIONS FOR FURTHER READING

Exploring Electricity. H. H. Skilling (Ronald Press).
A History of Electrical Engineering. P. Dunsheath (Faber & Faber).

Exercises

1.1. Make a dimensional analysis of the following quantities:

        (a) velocity,   (b) energy,
        (c) power,   (d) electric charge.

1.2. Compile a list of electrical words.

1.3. Distinguish between units and dimensions.

CHAPTER 2

Electric Charges at Rest (I)

2.1. THE FORCES BETWEEN ELECTRIC CHARGES

Most important electrical devices function because of the *motion* of electric charges, and many books therefore start with the discussion of electric currents. But it is not easy to understand the behaviour of moving charges until one is familiar with the action of electric charges at rest. This is very similar to the procedure in mechanics where it is helpful to discuss statical problems before turning the attention to the often more exciting problems of dynamics. This is not to say that static problems are unimportant; electrostatics has many important industrial applicaions: motor-car bodies, for instance, are painted by an electrostatic process.

Let us first discuss what we mean by electric charge, so that we shall have a firm basis from which to work. All matter consists of fundamental particles of which the most common ones are the neutron, the proton and the electron. Neutrons and protons form the nucleus of an atom and electrons move in orbits around the nucleus. The diameter of a nucleus is of the order of 10^{-14} m and the electron orbits, which determine the size of an atom, have diameters roughly 10,000 times that of the nucleus. The masses of the particles are (1) neutron 1.6747×10^{-27} kg, (2) proton 1.6724×10^{-27} kg, (3) electron 9.1083×10^{-31} kg. Thus the proton or neutron are 1,840 times as heavy as the electron.

Newton's law of gravitation states that there is a force of attraction between massive particles which varies inversely with

11

the square of the distance between the particles and can be written

$$F_g = \lambda_g \frac{m_1 m_2}{r^2} \qquad (2.1)$$

where λ_g is the constant of gravitation and m_1, m_2 are the masses. In the SI system of units λ_g is $6\cdot67 \times 10^{-11}$. If we calculate the forces between the three fundamental particles using eqn. (2.1) and compare our results with those deduced from experiment, we find some extraordinary discrepancies. The results are as follows:

(1) It is true that the law of force varies inversely as the square of the distance between the particles.

(2) Newton's law gives the correct answer for the force between neutrons, but not between electrons or protons.

(3) The forces between protons, between electrons, and between electrons and protons are quite different from those deduced from Newton's law. For one thing the forces are much bigger, for another they are in some cases repulsive instead of attractive forces.

We deduce from the experiments that protons and electrons have another property besides that of gravitational mass, and we call this property electric charge. Since the forces are attractive between electrons and protons, but repulsive between electrons and between protons, we postulate two types of electric charge. Moreover from the size of the forces we deduce that protons and electrons have equal amounts of charge, and from experiments on assemblages of protons and electrons in equal numbers we deduce that the effects of the two types of charge cancel. Hence we can describe the two types of charge as positive and negative. An arbitrary choice has to be made and the charge of the proton is taken as positive.

Every student of electrical engineering is likely to be asked by his friends to explain what electricity is. On the face of it this is a profound question, but generally the questioner is hoping for an explanation of electricity in terms of more familiar substances.

Such an explanation would be absurd because one would have to explain electricity in terms of aggregates of electrical charges, i.e. to explain a simple thing in terms of more complicated things. We have seen that electric charge like mass is a fundamental property of matter. A scientific answer, since science deals with measurement, would be to use the inverse square law of force as a definition: "Electric charge is that which when concentrated at a point will act on another point charge with a law of force inversely proportional to the square of the distance between the points." That is correct as far as it goes, but such an answer seldom carries conviction. Another possible answer is: "I do not know". This will also be true and it often gives a pleasing sensation to the questioner.

The inverse square law of force between charges is the foundation stone of electrical science. We can write it as

$$F_e = \lambda_e \frac{Q_1 Q_2}{r^2} \tag{2.2}$$

If Q_1 and Q_2 have the same sign there is repulsion, if they have opposite signs there is attraction. The unit of charge could have been chosen as that of a proton or electron; however, in the SI system the unit has been defined by an experiment with electric currents. This unit is called the coulomb and is immensely bigger than the electronic charge, which is $1 \cdot 60 \times 10^{-19}$ coulombs. From experiment we can determine the constant λ_e and it is found to be 9×10^9 (within 2 parts in 1000).

It is interesting to compare the electric and gravitational forces between two electrons. At a distance of 1 mm the gravitational force is an attraction of

$$F_g = 6 \cdot 67 \times 10^{-11} \times \frac{(9 \cdot 1 \times 10^{-31})^2}{(10^{-3})^2}$$

$$= 5 \cdot 53 \times 10^{-65} \text{ newtons} \tag{2.3}$$

The electric force is a repulsion of

$$F_e = 9 \times 10^9 \times \frac{(1 \cdot 6 \times 10^{-19})^2}{(10^{-3})^2}$$

$$= 2 \cdot 30 \times 10^{-22} \text{ newtons} \qquad (2.4)$$

Thus the ratio of the electric to the gravitational force is

$$F_e/F_g = 4 \cdot 17 \times 10^{42} \qquad (2.5)$$

Thus the electric force is immensely bigger than the gravitational force. Indeed, it is at first sight surprising that gravitational forces can be observed at all. The reason is that there are both positive and negative charges and that on average these cancel. Matter generally is electrically neutral, and electrical forces are balanced. In order to achieve electrical effects the electrical engineer has to separate positive and negative electric charges. Whereas electrical forces tend to cancel, gravitational forces are always attractive and additive. Thus it comes about that in spite of eqn. (2.5) motion on the earth and the motion of the heavenly bodies are controlled by gravitational and not by electric forces.

2.2. THE VERIFICATION OF THE INVERSE SQUARE LAW

A reader with a sense of history will suspect the argument of the last section, because the subject of electrostatics had developed long before there was any knowledge of neutrons, protons and electrons. He will also have some misgivings that in our discussion we have included gravitational and electrostatic forces but have left out the short-range forces which bind the particles together in the atomic nucleus. Moreover he might rightly be suspicious about the exact nature of the experiments which supposedly led to the results described.

It must at once be acknowledged that we have not followed the historical sequence. Many books on electricity have been written without reference to the fundamental particles, but since there are such particles there seems every reason to start our discussion by considering them. Nevertheless the laws of electromagnetism were originally applied to aggregates of electric particles and not to isolated particles or to nuclear structure. Additional considerations and laws have to be introduced when dealing with individual atoms. In this book we shall limit ourselves to the study of macroscopic phenomena and we shall treat electricity as a continuous fluid rather than as an assembly of discrete charges. This means that the smallest drop of our fluid must contain some thousands of electrons and the shortest distance between our charges must be about a thousand times the diameter of an atom. Thus distances of about 10^{-7} m are admissible and the restriction is hardly a severe one. Within this limitation we shall be able to speak of point charges, meaning by this term not a mathematical point of zero dimension but a region of 10^{-10} m diameter. Moreover the experiments that we shall cite in the construction of the theory of electromagnetism will deal with currents and charges containing large numbers of electrons.

The inverse square law is in fact as old as 1767, when Joseph Priestley suggested it. Priestley found that there was no electric force inside a metal cup which had been connected to an electric friction machine. This reminded him of a theorem of Newton's that inside a hollow sphere of gravitating matter there would be no gravitational force. Newton's experiment was an impossible one, at least at the time, because it would have had to be carried out in space away from other masses, but Priestley saw that his electrical experiment was strictly analogous; hence his suggestion of an inverse square law.

Priestley's friend Cavendish built an apparatus consisting of two concentric spherical conductors joined by a wire. The spheres were first charged, then the connecting wire was withdrawn and after the outer sphere had been earthed a pith-ball electroscope was used to explore the charge on the inner sphere. Cavendish found that there was no charge on this sphere and he concluded that, within his experimental accuracy, the force between charges must vary inversely as some power of the distance less than 2·02 and more than 1·98. A more recent experiment using an amplifier and galvanometer instead of an electroscope has shown that the index of the power law cannot differ from 2 by more than one part in 10^9. So our foundation stone of the inverse square law is well and truly laid.

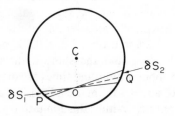

FIG. 2.1   *The force inside a hollow charged sphere*

The arguments underlying these experiments can best be understood with reference to Fig. 2.1. Consider the force at O, an arbitrary point within the hollow outer sphere. Draw a small conical surface through the apex O. This cone intersects the sphere in the two small areas δS_1 and δS_2 at P and Q. These two areas are proportional to the square of the distance from O, i.e. $\delta S_1 \propto OP^2$ and $\delta S_2 \propto OQ^2$ and therefore the amount of electric charge on δS_1 and δS_2 is proportional to the square of the distance from O. Hence if the forces from these two small areas are proportional to the charges on them and inversely proportional to the square of the distance from O, their effects cancel. By integration this is true for the whole spherical surface and zero force implies an inverse square law.

Curiously enough this law is generally called Coulomb's law after the great French experimenter and inventor of the torsion balance. His experiments were carried out some thirteen years after Cavendish, and they were also more difficult and less accurate. He had no knowledge of Cavendish's work because Cavendish did not publish his results.*

* He did, however, preserve his notebooks and these were edited by the great James Clark Maxwell about a century later.

2.3. THE ELECTRIC CONSTANT

The inverse square law of electrostatics states that the force between two charges is proportional to the product of the charges and inversely proportional to the square of the distance between them, i.e.

$$F_e \propto \frac{Q_1 Q_2}{r^2} \qquad (2.6)$$

or
$$F_e = \lambda_e \frac{Q_1 Q_2}{r^2} \qquad (2.2)$$

where λ_e is the constant of proportionality. The dimensions of this constant are Force (Length)^2/(Charge)^2 which is ML^3/T^4A in the M,L,T,A system of dimensions. The magnitude of the constant depends on the size of the units and in the SI system this is very nearly 9×10^9,

The original workers naturally thought that unity was much the best constant. Unfortunately this leads to a system of units in which electric current is measured in $1/(3 \times 10^9)$ amperes, as we have already noticed in Section 1.2. Thus the constant which has been banished from one relationship appears in different guise in another place. The reason is not far to seek and is of course a physical one. Electric forces are enormous and hence we shall need either a large constant of proportionality or a small unit of electric charge. Some of the protagonists of special sets of units seem to think that the laws of nature can be changed by international agreement, but engineers have to take the world as they find it. The electrical constant λ_e is defined once the units have been defined and it cannot be defined or measured independently from the measurement of electric charge.

One further complication has now to be faced before we can leave this basic but rather dreary discussion of units. This complication is a matter of geometry. In developing formulae for

EFE—C

electrical apparatus we shall often deal with spherical shapes, with cylinders and with rectangular boxes. Somewhere or other the factors 2π and 4π are bound to appear. It would be nice if 2π appeared in cylindrical problems and 4π in spherical ones, and we can ensure this by a cunning device recommended by Heaviside which he called *rationalization*. Instead of using λ_e as our constant in the inverse square law we shall use a constant which contains 4π explicitly. Since point charges refer to a spherical geometry, this will have the desired effect. Accordingly we shall write the inverse square law in the form

$$F = \frac{Q_1 Q_2}{4\pi\varepsilon_0 \, r^2} \qquad (2.7)$$

where ε_0 is now the constant instead of λ_e. There is no special reason for putting the constant into the denominator of the expression, but it is the general practice. The magnitude of ε_0 is $8 \cdot 854 \times 10^{-12}$ or very nearly $1/(36\pi \times 10^9)$.

Many writers call ε_0 the *permittivity of free space*, a term which suggests that this constant describes a property of space which could be measured independently. However, we have seen that ε_0 arises in the definition of electric charge through the inverse square law and that it is not an independent entity. There seems to be no reason for inventing unmeasurable properties of emptiness and in any case we shall need the term *permittivity* later, so we shall follow those authors who call ε_0 the *primary electric constant*.

2.4. THE ELECTRIC FORCE IS A VECTOR QUANTITY

In order to define a force acting on a body we have to know its magnitude and direction. Force is a vector quantity and the force between electric charges is no exception to this rule. When we are dealing with two point charges by themselves then this force must act along the line joining the charges. This is a consequence of Newton's third law that action and reaction are equal and

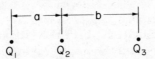

FIG. 2.2   *Three charges in a line*

opposite. When there are several charges we must add the forces vectorially. Consider two simple cases. In Fig. 2.2 three positive charges lie along the same line. The force on Q_2 from left to right is given by:

$$F = \frac{Q_1 Q_2}{4\pi\varepsilon_0 \, a^2} - \frac{Q_3 Q_2}{4\pi\varepsilon_0 \, b^2}$$

$$= \frac{Q_2}{4\pi\varepsilon_0}\left(\frac{Q_1}{a^2} - \frac{Q_3}{b^2}\right) \tag{2.8}$$

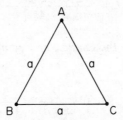

FIG. 2.3   *Three equal charges at the corners of a triangle*

In Fig. 2.3, three equal charges lie at the corners of an equilateral triangle ABC of side a. The force on the charge Q at A is given by the vector sum of a force

$$F = \frac{Q^2}{4\pi\varepsilon_0 \, a^2} \tag{2.9}$$

along BA and a similar force along CA. Thus the resultant force on Q at A is

$$F = 2\sin 60° \frac{Q^2}{4\pi\varepsilon_0 \, a^2} = \frac{\sqrt{3}.Q^2}{4\pi\varepsilon_0 \, a^2} \tag{2.10}$$

perpendicular to BC.

2.5. ELECTRIC FIELD STRENGTH

Let us now take a close look at eqn. (2.8). There are two forces, which both contain Q_2 and this is reasonable because we are concerned with the force on Q_2. In the general case $F = Q_2 \times$ function $(Q_1, Q_3, Q_4, \ldots, Q_n)$. We can thus take Q_2 out as a common factor and this suggests another way of looking at the problem. Instead of applying the inverse square law directly we can proceed in two steps as follows.

The force on Q_2 is proportional to Q_2, let it be written as EQ_2, where E describes the effect of the charges Q_1 and Q_3 at the place where Q_2 is. In the case of Fig. 2.2

$$E = \frac{Q_1}{4\pi\varepsilon_0 \, a^2} - \frac{Q_3}{4\pi\varepsilon_0 \, b^2} \tag{2.11}$$

along the line joining the charges and acting from left to right. In the case of Fig. 2.3

$$E = 2\sin 60° \frac{Q}{4\pi\varepsilon_0 \, a^2} = \frac{\sqrt{3} \cdot Q}{4\pi\varepsilon_0 \, a^2} \tag{2.12}$$

perpendicular to BC.

E has magnitude and direction and is called the *electric field strength*. Its dimensions are force divided by charge: ML/T^3A. At every point in space in the neighbourhood of an electric charge Q we can calculate the electric field strength, which is simply

$$E = \frac{Q}{4\pi\varepsilon_0 \, r^2} \tag{2.13}$$

where r is the distance from Q. If the charge is positive, E is outward from Q and if the charge is negative E is towards Q. The vector E describes the *electric field*. Thus we can now say that charge A produces the field E_A which acts on charge B, whereas previously we should have said charge A acts on charge B. The invention of E does not rest on any new experimental evidence, since the field cannot be observed unless there is a charge in it which can experience a force. But the idea of a field is of tremendous help in solving problems, because it allows us to think of the charges one at a time. Moreover sometimes there may be an obstacle between the charges, for instance in Fig. 2.2 there might be a curtain between Q_2 and Q_3. Then it would be impossible to know what charges were acting on Q_2, but we could still observe the electric field strength at Q_2. In many cases of engineering interest the sources of the field are inaccessible, we therefore turn our attention to the field rather than to the charges which cause the field. The notion of a field is due to Gilbert. At the time of Franklin it was thought that the idea of electric charge had proved the field idea to be erroneous, but we now see that what is really needed is a composite view which makes use of both charge and field.

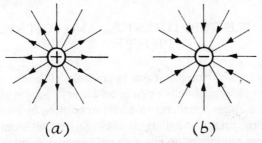

(a) (b)

Fig. 2.4   (a) Electric field of a positive point charge
(b) Electric field of a negative point charge

If it is required to find the direction of the electric field strength at a large number of points, it becomes worth while to construct a map. The two simplest maps are those of a point positive charge and a point negative charge as shown in Fig. 2.4. It must of course be remembered that the map only shows a two-dimensional section through the three-dimensional field. The actual field of a point charge is like a hedgehog with bristles sticking out in all directions.

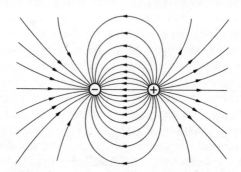

FIG. 2.5   *Field of two equal and opposite point charges*

A more interesting map is one showing the field directions near two equal and opposite point charges. This is given in Fig. 2.5.

2.6. ELECTRIC POTENTIAL AND POTENTIAL DIFFERENCE

Any system of charges at rest is unstable. The inverse square law causes charges of unlike sign to collide and charges of like sign to separate unless the charges are held in position by forces which are not electrostatic. Similarly the heavenly bodies would collide if they were not in motion but were influenced only by the inverse square law of gravitational attraction. In the same way a marble rolls down a smooth plane unless it is held in position by some force, which is not gravitational.

Work has to be done to assemble systems of charges and this work can be recovered when they are released. Thus the systems possess potential energy.

To fix our ideas let us consider the potential energy of two charges Q_1 and Q_2 of like sign at a distance a from one another

(Fig. 2.6). What is the work that has to be done to assemble this system? First let us place in position the charge Q_1 at A. This requires no work because it is the only charge present and so it

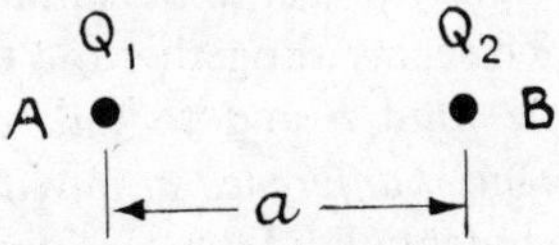

$$Q_1 \qquad\qquad Q_2$$
$$A \bullet \qquad\qquad \bullet B$$
$$|\!\leftarrow\!\!-\!\! a \!-\!\!\rightarrow\!|$$

FIG. 2.6　*Two charges at a distance* a *from one another*

experiences no force. Now let us carry the second charge Q_2 towards B. This charge experiences a force

$$F = \frac{Q_1 Q_2}{4\pi\varepsilon_0 r^2} \text{ at a distance } r \text{ from A}$$

The work required will be

$$\int_\infty^a -F\,dr = \int_a^\infty F\,dr = \frac{Q_1 Q_2}{4\pi\varepsilon_0}\left[-\frac{1}{r}\right]_a^\infty$$

$$= \frac{Q_1 Q_2}{4\pi\varepsilon_0 a} \qquad\qquad (2.14)$$

The limits of the integral imply that the charge Q_2 is brought from an infinite distance and that the negative sign in front of F is put there to show that work is done against the force. Equation (2.14) gives the potential energy of the system.

Another way of obtaining this result is to make use of the electric field strength E. The potential energy is then given by

$$\int_\infty^a -EQ_2\,dr = Q_2\int_a^\infty E\,dr = \frac{Q_2 Q_1}{4\pi\varepsilon_0 a} \qquad (2.15)$$

The expression

$$\int_\infty^a -E\,dr = \frac{Q_1}{4\pi\varepsilon_0 a} \qquad\qquad (2.16)$$

gives the potential energy divided by the charge Q_2. It is thus the potential energy per unit charge and it is called the *potential* of

Q_2.* Its unit is the joule/coulomb or volt and its usual symbol is V. It is a scalar and not a vector quantity, since energy has magnitude but not direction.

Some readers will rightly object to the limit of the integral being taken at infinity, which seems altogether too theoretical. A better choice is to set some limit b and to talk not of the potential of Q_2, but of the *potential difference* in moving from a distance b to a distance a. This potential difference is given by the expression

$$V = \int_b^a -E\,dr = \frac{Q_1}{4\pi\varepsilon_0}\left[\frac{1}{a} - \frac{1}{b}\right] \qquad (2.17)$$

The expression *potential difference* is often written p.d.

In eqn. (2.14)–(2.17) we have obtained the potential, or potential difference, by integration. Clearly the process could be reversed and we could obtain the electric field strength by differentiation of the potential. The electric field strength is the *gradient* of the potential and its unit is the volt/metre. Of course this is familiar ground from our knowledge of hills and mountains. Potential corresponds to height above sea level and electric field strength corresponds to gradient or line of steepest descent. This correspondence between a mountainous country and a region containing electric charges is a direct consequence of the fact that the inverse square law governs both gravitational and electrostatic problems and it suggests at once that the electric field can be described graphically by means of a contour map. The contours in the electrical case are lines of equal potential above some arbitrary datum and are generally called equipotentials. In a three dimensional model the equipotentials are not lines but surfaces. Figure 2.7 shows the equipotentials around a point charge for equal steps of potential. The unequal spacing is of course due to the fact that the expression for the potential at a distance r from the charge is given by

$$V = \frac{Q}{4\pi\varepsilon_0\,r} \qquad (2.18)$$

* It should be noted that although the potential is said to belong to Q_2, it is actually a property of the system Q_1 and Q_2.

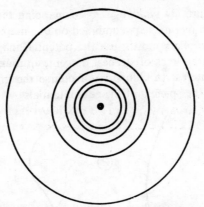

FIG. 2.7   *Equipotentials of a point charge*

Figure 2.7 should be compared with Fig. 2.4 which describes the same problem. Figure 2.8 shows a graph of V plotted against r, which is a rectangular hyperbola. It will be seen that the potential increases very steeply the nearer the second charge approaches to

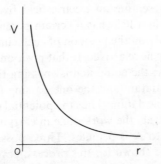

FIG. 2.8   *Potential of a point charge*

the charge Q. This is to be expected because the force varies as the inverse square of the distance. Thus both force and potential would reach infinite values if the second charge could be placed exactly on top of the charge Q, but as we have shown in Section 2.2, the theory only holds for distances down to 10^{-7} m.

Clearly there are always two ways of mapping the electric field and often both methods are combined on one map. It is interesting that geographers mostly use the potential map and not the map showing the lines of steepest descent, although these might be of great interest to climbers. Of course the gradient can be inferred from the spacing of the equipotentials.

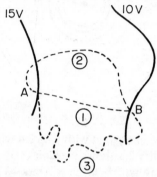

FIG. 2.9   *Work done in moving a charge between two equipotentials*

There is one very important conclusion that follows from the possibility of representing an electric field by a potential map. Since every point in a field has a certain potential, the potential energy depends only on the position of each charge and not on the path by which they have arrived at that position. Figure 2.9 shows two equipotentials; the equipotential passing through A is at 15 volts above some datum and the one passing through B is at 10 volts above the same datum. Thus the potential difference between A and B is 5 volts, i.e. the work done in carrying a charge of one coulomb from B to A is 5 joules. Three possible paths of very different length are shown for this process, nevertheless the work done is the same in each case because it depends only on the final position. Moreover if the charge is carried from B to A by one path and returns to B by the same or any other path, no work will have been done. All the work expended will have been recovered. A system for which this is true is called a *conservative* system and any system of forces for which one can draw a potential map is a conservative system.

All this is fairly obvious until one realizes with a shock that in mountaineering it is clean contrary to experience. Surely it is quite absurd to say that one has done no work in climbing a mountain and coming down again, nor is the amount of energy expended independent of the path taken. It is of tremendous importance that we should face and resolve this difficulty.

The answer is as follows. The contour or potential map shows only gravity forces and these forces by themselves form a conservative system. But a mountaineer also experiences other forces, chiefly frictional ones, and these convert kinetic and potential energy into heat, which is dissipated and lost. Frictional forces are called dissipative or non-conservative and they depend on the path chosen by the climber. In electricity also there are non-conservative forces, but the inverse square law of electrostatics by itself gives rise to a conservative system. Non-conservative forces in electricity are associated with the motion and particularly the rapidly accelerated motion of electric charges. Great caution is needed when talking about potential difference in systems where electric charges are in motion. In some cases the electrostatic forces predominate, then all is well. But if they do not, as for instance in a radio aerial, then any talk about potential difference tends to be nonsense.

One other point needs to be stressed. If the charges in an electrostatic system are slowly moved round a complete cycle, no work is done. It is therefore not possible to abstract energy from such a system by means of a continuously repeated process. This is not surprising; after all, the energy must come from somewhere. Nevertheless it is a point overlooked by some hopeful inventors. Electrostatic systems are no use for energy conversion, but they are exceedingly useful for energy storage, as we shall see in the next chapter.

2.7. CONDUCTORS IN ELECTROSTATIC FIELDS

A conductor is a material through which electric charge can move easily. There are gaseous conductors and liquid conductors, but in this book we shall generally be thinking of solid conductors and particularly of metals. In an uncharged piece of metal there are equal numbers of protons and electrons. Some of the electrons are free to move and they therefore distribute themselves uniformly throughout the volume until every part of the material is electrically neutral and there are no electric forces to cause further motion.

If the metal is charged by adding electrons or removing some of the existing ones, it can be regarded as a neutral body with additional negative or positive charges. Because of the inverse square law the additional charges will separate as far from one another as they can, and after a very short time all the charges will be on the surface of the metal.

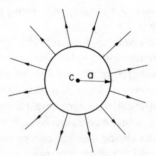

FIG. 2.10   *Field of a charged conducting sphere*

Inside the metal there is no electric field strength. If there were, the electrons would move, but we are discussing the static case when all motion has ceased. Since there is no electric field strength inside the metal, there are no differences of potential energy in the metal and the body is at a uniform potential. Thus in electro-statics all conductors are regions of uniform potential which is also the potential of the surface.

So far we have dealt only with point charges, but now we can extend the treatment to charged conductors. It is clear that the field outside the surface will not be disturbed if we replace an equipotential surface by a metal surface at the same potential. Consider the simple case described in Fig. 2.10. The potential at a distance a from the charge Q is $Q/4\pi\varepsilon_0 a$. Thus the field will not be disturbed if we replace the equipotential at a by a con-ducting spherical surface at a potential $Q/4\pi\varepsilon_0 a$. The external field for radii greater than a is thus either the field of a point charge or that of a charged sphere at the potential $Q/4\pi\varepsilon_0 a$. The problem of the sphere of any radius can therefore be derived from the case of a point charge.

SUMMARY

In electricity the foundations are the most difficult part of the subject and this chapter has been difficult. We have defined the property of electric charge in terms of the inverse square law of force. The constant of proportionality in this law has been considered and we have seen that its magnitude depends on the unit of electric charge. We have invented the term *electric field strength* to simplify the calculation of forces on electric charges and we have shown that the idea of an electric field is especially useful when some of the charges are inaccessible.

We have calculated the work that has to be done in assembling two electric charges and have considered the potential energy of electrical systems. This has led to the definition of electric potential and potential·difference. We have shown that an electrostatic field can be described by drawing a contour map, in which the contours are equipotentials. We have also shown that an electrostatic system is conservative and that its chief use is to store energy.

Finally we have had a brief look at the behaviour of conductors in electrostatic fields and have found that in charged conductors the charge always forms a surface layer and that the potential of a conductor is the same all over it.

TERMS USED IN THIS CHAPTER

Term	Symbol	Unit (Abbreviation)	Definition
Gravitational Mass	m	kilogram kg	$F = \lambda_g \dfrac{m_1 m_2}{r^2}$
Electric Charge	Q	coulomb C	$F = \dfrac{Q_1 Q_2}{4\pi\varepsilon_0 r^2}$
Electric Field Strength	E	volt/metre V/m	$E = \dfrac{Q}{4\pi\varepsilon_0 r^2}$
Potential	V	volt V	$V = \displaystyle\int^r - E\,\mathrm{d}r$
Potential Difference	V	volt V	$V = \displaystyle\int_1^2 - E\,\mathrm{d}r$

Exercises

2.1.   Define the term *electric field strength*. What is its unit in the SI system, and what are its dimensions?

2.2.   What would be the force between two charges, each of one coulomb, if it were possible to place them at a distance of 10 mm from each other? (*Ans.* 9×10^{13} newtons)

2.3.   What is the potential energy of the system described in Ex. 2, if both charges are of equal sign? (*Ans.* 9×10^{11} joules)

2.4.   Sketch the potential map of two equal charges separated from one another by a distance a.

2.5.   Sketch the lines of electric field strength for the arrangement of Ex. 4.

2.6. Explain the term *electric potential*. State the unit in which it is measured in the SI system and show the relationship of this unit to the ampere.

2.7. The specification for an electric motor shows that the device consists in essence of an arrangement of charged conductors in relative motion. You are asked to advise a firm, who are considering the manufacture of the motor. Embody your advice in a suitable letter.

2.8. Explain why lines of electric field strength are always perpendicular to equipotentials.

2.9. Two conducting balls of radius 0·1 m are situated 3 m apart in free space. Electrons are transferred from one to the other at a rate of 10^{13} per second. Estimate the time taken for the balls to develop a p.d. of 100 kV.

(*Ans.* 0·36 sec)

CHAPTER 3

Electric Charges at Rest (II)

3.1 THE ELECTRIC FIELD AS A MEASURE OF QUANTITY

The idea of the electric field enables us to examine the space in the vicinity of electric charges without having to examine the charges themselves. Thus if the electric field strength in a place is E volts/metre, we know that the force on a charge Q coulombs at that point will be EQ newtons in the direction of E. We do not know whether E is caused by a single charge or by some arrangement of charges and very often we do not need to know the sources of E. Similarly, if it is specified that there is a potential difference of V volts between two points, then we know that the change of potential energy as a charge Q coulomb is moved between these points is VQ joules. We do not know, and often do not care, what has caused the potential difference.

The usefulness of the electric field and of the quantities E and V is thus firmly established and one might leave it at that. But consideration of such field maps as Figs. 2.4 and 2.5, pages 21 and 22, suggests that there may be even more use in the electric field than is implied by electric field strength and potential difference.

Consider Fig. 2.4. It is clear that the charges are centres of the field, the lines of E diverge from (or converge on) the charges. Suppose the charge is inaccessible because there is a fence in the form of a closed surface of arbitrary shape around it (see Fig. 3.1). Surely one ought to be able to obtain some information about the charge inside the fence by examining the electric field strength outside the fence. If E is dominantly outwards it seems likely that there is positive charge inside, and if E is dominantly inwards there must be negative charge inside.

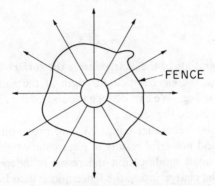

FIG. 3.1 *The field of an inaccessible charge*

Let us try this suggestion by examining a very simple case. Figure 3.2 shows a point charge surrounded at a radius a by a

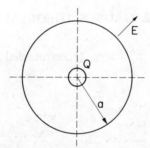

FIG. 3.2 *Point charge surrounded by a spherical fence*

fence in the form of a spherical shell. Just outside the fence the electric field strength is given by

$$E = \frac{Q}{4\pi\varepsilon_0 a^2}$$ (3.1)

EFE—D

By symmetry the direction of E is radially outwards. The surface area of the fence is

$$S = 4\pi a^2 \qquad (3.2)$$

and we notice that

$$Q = \varepsilon_0\, ES \qquad (3.3)$$

Thus the magnitude of the charge inside the surface is equal to the product of ε_0, the electric field strength at the surface, and the area of the surface. We can deduce the size of the charge without having to go inside the fence.

We shall now generalize eqn. (3.3) and prove one of the most remarkable and powerful results of electrostatics. We shall show that with a small modification the result is independent of the position of the charge inside the fence and is also independent of the shape of the fence. In the proof we shall make use of the inverse square law only, and the theorem, which is due to Gauss, can therefore be applied to those other branches of science in which the inverse square law holds.

3.2. GAUSS'S THEOREM

Figure 3.3 shows a charge Q surrounded by a surface of arbitrary shape.

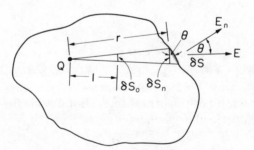

FIG. 3.3   *Charge surrounded by arbitrary surface*

Consider the product of ε_0, the electric field strength and the area at a point on the surface.

$$E = \frac{Q}{4\pi\varepsilon_0 \, r^2} \tag{3.4}$$

The component of E normal to the surface is

$$E_n = E \cos \theta \tag{3.5}$$

The product of ε_0, E_n and the surface area is

$$\varepsilon_0 \, E_n \, \delta S = \varepsilon_0 \, E \cos \theta \, \delta S = \varepsilon_0 \, E \, \delta S_n$$

$$= \frac{Q}{4\pi} \frac{\delta S_n}{r^2} \tag{3.6}$$

Now δS_n forms the base of a small cone with its vertex at Q.

Hence
$$\frac{\delta S_n}{r^2} = \frac{\delta S_0}{1^2} \tag{3.7}$$

and
$$\varepsilon_0 \, E_n \, \delta S = \frac{Q}{4\pi} \delta S_0 \tag{3.8}$$

If we now add all the possible terms of eqn. (3.8) by integration we have

$$\iint_S \varepsilon_0 \, E_n \, \mathrm{d}S = \frac{Q}{4\pi} \iint_{S_0} \mathrm{d}S_0 \,^*$$

Now $\iint_{S_0} \mathrm{d}S_0 = 4\pi$, the surface area of a sphere of unit radius.

Hence
$$\iint_S \varepsilon_0 E_n \mathrm{d}S = Q \tag{3.9}$$

Thus the charge inside is equal to the surface integral of $\varepsilon_0 E$. Also, since this is true for any charge Q, it is true for the sum of all such charges, wherever they are inside the surface.

Let us apply Gauss's theorem to some problems of finding the electric fields of charged bodies.

---

* The double integral sign reminds us that the integration is over a surface The integration of $\mathrm{d}S_0$ is 4π because it is a sphere of unit radius.

3.2.1. *The Field of a Uniformly Charged Sphere*

Consider a charged conducting sphere of radius *a* (Fig. 3.4). The charge will be concentrated in a surface layer and by symmetry the charge density will be uniform over the surface. By

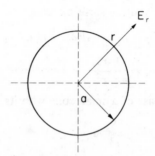

FIG. 3.4   *Field of a uniformly charged sphere*

symmetry the electric field outside the sphere will be in a radial direction and will be constant at any particular distance from the sphere. By Gauss's theorem,

$$\iint_S \varepsilon_0 E_n \, dS = Q \tag{3.9}$$

Therefore, at radius *r*:

$$\varepsilon_0 E_r 4\pi r^2 = Q$$

and

$$E_r = \frac{Q}{4\pi\varepsilon_0 r^2} \tag{3.10}$$

Thus the field of a charged sphere is the same as if the charge were concentrated at the centre of the sphere. This is a conclusion we had almost reached in Section 2.7. There, however, we were only thinking of the potential: the equality of the charges is a result of Gauss's theorem.

3.2.2. *The Field Close to the Surface of a Charged Conductor*

The surface which has a charge density of q coulomb/m^2 is shown in Fig. 3.5. It is a flat surface, but by this we merely mean that very close to the surface it looks flat in the same sort of way

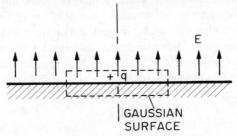

FIG. 3.5 *Field close to surface of a conductor*

that the earth appears flat to us. We shall apply Gauss's theorem to a shallow cylindrical box closed by the "Gaussian surface" shown dotted in the figure.

Consider
$$\iint_S \varepsilon_0 E_n \, dS = Q \tag{3.9}$$

There is no electric field inside the conductor and hence no contribution to the integral there. By symmetry there is no E_n at the curved part of the box.

Hence
$$\iint_S \varepsilon_0 E_n \, dS = \iint_S \varepsilon_0 E \, dS \tag{3.11}$$

also
$$Q = \iint_S q \, dS \tag{3.12}$$

Therefore
$$\varepsilon_0 E = q \tag{3.13}$$

and
$$E = \frac{q}{\varepsilon_0} \tag{3.14}$$

Thus the electric field strength at a point just outside the surface of a charged conductor is equal to the surface density of charge at that point divided by the electric constant.

3.3. ELECTRIC FLUX DENSITY AND ELECTRIC FLUX

Our investigations with Gauss's theorem have shown that the electric field can give information not only about the forces on charges, but also about the magnitude of the (source) charges which cause these forces to exist. Of course both these bits of information are aspects of the same phenomenon, but it is worth keeping the two aspects separately in view. This thought leads to the invention of a new symbol D, where

$$D = \varepsilon_0 E \tag{3.15}$$

In Section 3.2.2. we have seen that at the surface of a conductor

$$D = q \tag{3.16}$$

where q is the charge density in coulomb/m^2. D is given the name *electric flux density*, and

$$\Psi = \iint D_n \, dS \tag{3.17}$$

is called the electric flux. Gauss's theorem states that for a closed surface

$$\Psi = Q \tag{3.18}$$

The idea behind the terms *flux* and *flux density* is that the effect of the charges is conveyed to a distant place by something like a flux. This is a bit misleading, because there is no flow of anything. All we know is that charges act on one another at a distance with an inverse square law of force. However, the terms flux and flux density are well established and it is a waste of time to argue about names. The crux of the matter is that D and Ψ are useful to us, because they embody Gauss's result that over an area remote from a charge we can gain information about the size of that charge. The flux Ψ over a closed surface is equal to the charge enclosed and the flux density D will give information about the charge density on a surface.

3.4. TUBES OF FLUX

When we were discussing the idea of potential difference, we found it useful to draw a potential map, which showed a number of equipotentials at fixed intervals of potential. Such a map shows not only the total difference in potential between two places, but also the manner in which the potential varies. The question

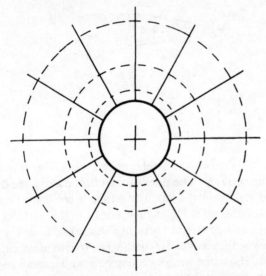

FIG. 3.6 *Field of a uniformly charged sphere*

naturally arises whether in a map showing the electric field strength, and therefore the electric flux density, one cannot also show how the field varies in strength from place to place.

Consider the case of a uniformly charged sphere illustrated in Fig. 3.6. Close to the sphere the equipotentials crowd together and so do the lines of flux density; it is clearly here that the field is strongest. Of course this crowding together arises from the fact that we have chosen equal steps of potential and equal angles between the lines of flux density. In the symmetrical case of Fig. 3.6 this is reasonable.

Now let us make it a rule that we shall not only divide up the potential into equal steps, but also the flux. Since we are dealing with three-dimensional problems this means that we shall draw equipotentials as surfaces and flux as tubes.

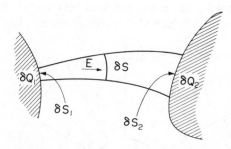

FIG. 3.7   *A tube of flux*

Figure 3.7 shows such a tube of flux extending between two conducting surfaces. The sides of the tube lie along the direction of E (and D). Thus the flux in the tube is constant, because no flux crosses the sides. The tube ends on the surfaces of the conductors and intersects these in the areas δS_1 and δS_2. On these areas lie the charges δQ_1 and δQ_2. We are using differential notation to show that we are dealing only with a small portion of the conductors A and B.

Since the flux is constant along the tube and since flux is associated with charge, it is reasonable to expect that the charges at the ends must be equal in magnitude. They are in fact equal in magnitude and of opposite sign, as can be shown by Gauss's theorem. Consider in Fig. 3.8 a Gaussian surface formed by the tube of flux and two small extensions into each conductor. No flux crosses this surface, which therefore must contain no net charge. Thus

$$\delta Q_1 + \delta Q_2 = 0$$

$$\delta Q_1 = -\delta Q_2 \tag{3.19}$$

Thus any tube of charge ends on charges of equal magnitude and opposite sign. By splitting up the field into tubes of flux we

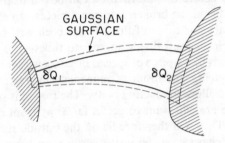

FIG. 3.8   *The charge at the ends of a tube of flux*

associate each tube with a definite amount of charge, and a tube of flux can end only on a charge. If all the tubes end on equal amounts of charge, we can calculate the total charge by counting the tubes of flux.

The electric field strength at any point can be obtained by dividing the flux in the tube by the cross-sectional area of the tube and ε_0 (see Fig. 3.7).

Hence $$E = \frac{1}{\varepsilon_0} \frac{\delta \Psi}{\delta S} \tag{3.20}$$

This follows from the definition of

$$\Psi = \iint D_n \, dS \tag{3.17}$$

and $$D = \varepsilon_0 E \tag{3.15}$$

This discussion of tubes of flux seems to most students a monstrous example of perverted ingenuity. Yet curiously enough the idea forms one of the most useful weapons in the armoury of an electrical engineer. If all our apparatus consisted of beautifully symmetrical bodies such as spheres and cylinders, we could solve our problems by analysis, using Gauss's theorem and symmetry. But generally the apparatus is of varying shape and we have to make use of numerical methods rather than neat equations. Happily, computers can be used to take the labour out of such methods, but before the information can be fed to the computer the problem has to be broken into small pieces. In electrostatics these pieces are our tubes of flux. It has been well said that the electrical engineer must divide space into tubes—not into cubes!

If this sounds like a piece of special pleading we shall strengthen the case by giving one result of immediate interest: there is no field inside a hollow conducting body. The reason is simple. Any charges there are will want to get as far away from one another as possible. They will therefore lie on the outside surface. Since there are no charges on the inside surface, no tubes of flux can start from the inside surface and where there is no flux there is no field. Thus the proposition is proved.

3.5. THE STORAGE OF ELECTRIC ENERGY

In our discussion of potential and potential difference we have already mentioned that assemblages of electric charges possess potential energy. Work has to be done to assemble the charges and this work can be recovered when the charges return to their original positions. Electrostatic systems can, therefore, be used as stores of energy and this in fact is their chief use in engineering.

The amount of energy that can be stored in this manner is small, the reason for this is that the forces between charges are enormous and if one tries to assemble large amounts of charge there is likely to be a flashover. This of course is what happens in a thunderstorm. Where it is necessary to store large amounts of energy, engineers make use of the potential energy of water pumped to high reservoirs, or the kinetic energy of flywheels or of electric currents.

There are, however, two ways in which electrostatic devices are supreme. The first is that they are wonderfully efficient. Energy losses as low as 1 in 10^5 are quite possible. The second advantage is that electrostatic devices have very little inertia. In mechanical devices protons and neutrons contribute to the inertia, but in electrical devices only electrons need to be moved. In high speed applications, where the time is measured in microseconds, electrostatic systems are without rival.

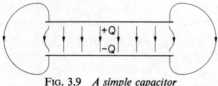

FIG. 3.9   *A simple capacitor*

We must now consider how to develop suitable "packages" for storing electric energy. First there is the choice between systems of like charges and those of unlike charges. Since bodies are normally electrically neutral it is easier to use unlike charges. An electric battery, for instance, will produce unlike charges at its terminals. The second question concerns the shape of the package. We know that the energy is large when the charges are close together, so a convenient shape would be to have closely spaced metal plates. The simplest case is that of Fig. 3.9 which shows two metal plates carrying electric charges $+Q$ and $-Q$. Such an arrangement is called a capacitor, because of its capacity for storing electric charge and energy.

3.6. CAPACITANCE

We now need a convenient measure by which capacitors may be compared. Suppose a potential difference V is maintained between the plates of a capacitor, possibly by means of a battery. This potential difference results in charging one plate positively and the other negatively with a charge Q (see Fig. 3.9). Now let the potential difference be doubled, possibly by putting two batteries in series. This will mean that the electric field strength will be doubled and therefore the flux density will be doubled. But the flux density is a measure of the charge density and this will also be doubled. Hence Q will be doubled. It is clear that the charge on the plates is proportional to the potential difference between them. Hence a suitable measure which is independent of voltage and charge is the ratio

$$C = \frac{Q}{V} \qquad (3.21)$$

C is called the *capacitance* and its unit is the farad. One farad is thus one coulomb/volt.

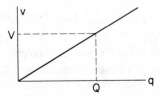

Fig. 3.10 *Relationship between charge and voltage*

The energy stored in the capacitor is proportional to the charge Q and the potential difference V. It is tempting to write Energy = Charge × Voltage, but this is wrong. Consider Fig. 3.10 which shows the relationship between charge and voltage. Let the voltage be v at any instant and let a small amount of charge δq be transferred at this voltage. The work done will be $v\,\delta q$. Hence the total work will be

$$W = \int_0^Q v\,\mathrm{d}q = \int_0^Q \frac{q}{C}\,\mathrm{d}q$$

$$= \frac{1}{2}\frac{Q^2}{C} \qquad (3.22)$$

Thus the work done, or the energy stored is given by

$$W = \frac{1}{2}\frac{Q^2}{C} = \frac{1}{2}CV^2 = \frac{1}{2}QV \qquad (3.23)$$

3.6.1. *Capacitance of a Parallel-Plate Capacitor*

We shall neglect for the moment the fringing part of the field in Fig. 3.9 and take instead the idealized field of Fig. 3.11. Let

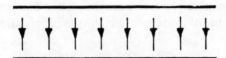

FIG. 3.11   *Capacitor without fringing flux*

there be a charge Q on each plate. The electric flux density will be given at the surface of the plate by

$$D = \frac{Q}{S} \qquad (3.24)$$

where S is the area of each plate. Since the field is uniform, the value of D will be the same everywhere between the plates. Hence the value of E everywhere will be

$$E = \frac{Q}{\varepsilon_0 S} \qquad (3.25)$$

The potential difference is the work done in carrying a unit charge from one plate to the other

$$V = \int -E\,\mathrm{d}l = -Ed$$

$$= \frac{Q}{\varepsilon_0 S}d \qquad (3.26)$$

where d is the distance between the plates.

Hence
$$C = \frac{Q}{V} = \frac{\varepsilon_0 S}{d} \tag{3.27}$$

.   Thus the capacitance is proportional to the area of the plates and inversely proportional to the distance between the plates.

The energy stored in the capacitor could be altered by moving the plates and thus altering the spacing. But this would produce a system with large inertia and a much simpler method is to leave the plates fixed but vary the charge on the plates. Only electrons have to be moved and even these form only a small fraction of the total number of electrons in the plates. This is the secret of the universal use of capacitors in high-speed applications.

3.6.2. Capacitance of a Cylindrical Capacitor

The arrangement is shown in Fig. 3.12. The radius of the inner conductor is b and the inner radius of the outer conductor is a. Let there be a charge $+Q$ on the inner conductor and $-Q$ on the

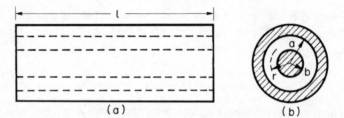

(a)   (b)

FIG. 3.12   *Cylindrical capacitor*

outer conductor and let l be the length of the capacitor. At any radius r the electric flux density can be obtained by Gauss's theorem

$$D \times 2\pi r l = Q \tag{3.28}$$

Hence
$$E = \frac{Q}{2\pi\varepsilon_0\, rl} \qquad (3.29)$$

and
$$V = \int_a^b -E\,\mathrm{d}r$$

$$= \frac{Q}{2\pi\varepsilon_0 l}\, \ln(a/b) \qquad (3.30)$$

Hence
$$C = \frac{Q}{V} = \frac{2\pi\varepsilon_0\, l}{\ln(a/b)} \qquad (3.31)$$

Note that in eqn. (3.30) a is greater than b. We have obtained a positive value for V by integrating from the radius a to the radius b. This is a reminder that the inner conductor is at the higher potential. This of course is correct because the charge on it is positive.

3.7. INSULATING MATERIALS BETWEEN THE PLATES OF CAPACITORS

In discussing capacitance we have so far ignored the material between the conductors. If the material is air this does not matter, because the capacitance in air is much the same as in a vacuum. However, it is well known that the capacitance can be greatly increased by inserting an insulating material, such as mica, between the plates. Since the plates have to be kept apart by some means, such an insulating material, or dielectric, serves the double purpose of separating the plates and increasing the capacitance. Let us examine how this increase in capacitance comes about.

In an insulator the electrons are bound to the atoms and are not free to wander through the material under the action of an electric field, as they are in a conductor. Nevertheless there can be a displacement of electric charge, which comes about as follows.

Figure 3.13 shows a model of an atom consisting of a positively charged nucleus and a system of electrons in orbit around the nucleus. In the first part of the figure there is no applied electric field and the atom also is electrically neutral, but in the second part of the figure the applied field has distorted the atom which is

(a) (b)

FIG. 3.13   *Polarization of an atom*

now equivalent to a small doublet (or dipole) of charge oriented along the electric field. Inside the material many atoms overlap, so that on average there is no noticeable effect, but at the edges of the material surface layers of charge appear as shown in Fig. 3.14.

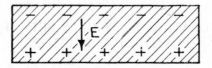

FIG. 3.14   *A polarized dielectric*

The effect is thus much the same as if there were free charges in the material, but the amount of surface charge is always less in an insulator than in a conductor. because in an insulator an electric field E is necessary to maintain the displacement. In a conductor the charges would continue to move until the field E had been cancelled entirely by the field of the surface charges.

The model of Fig. 3.13 is accurate as far as most dielectrics are concerned. Some materials, however, have *permanent* electric dipoles. The action of the electric field is then to align the dipoles which originally point in all directions. The resulting effect is again that shown in Fig. 3.14.

Consider now a parallel-plate capacitor with an insulating material between its plates (Fig. 3.15). Let Q be the charge on the

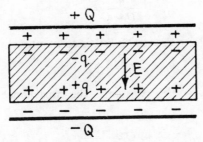

FIG. 3.15   *Parallel-plate capacitor with dielectric*

metal plates and q the induced surface charge on the insulator. The electric field between the plates is now due to Q and q.

Hence
$$E = \frac{Q-q}{\varepsilon_0 S}$$
(3.32)

where S is the area of the plates.

Hence
$$V = \int -E \, dl$$

$$= \frac{Q-q}{\varepsilon_0 S} d$$
(3.33)

where d is the distance between the plates,

and
$$C = \frac{Q}{V} = \frac{Q}{Q-q} \frac{\varepsilon_0 S}{d}$$
(3.34)

Comparing eqn. (3.27) and (3.34) we see that the capacitance has been increased in the ratio

$$\varepsilon_r = \frac{Q}{Q-q}$$
(3.35)

where ε_r is called the *permittivity* of the insulating material.

EFE—E

The amount of charge displacement q is in many materials proportional to the applied field E, which itself is proportional to $Q - q$.

Thus
$$\frac{q}{Q-q} = \chi_e \qquad (3.36)$$

where χ_e, the electric *susceptibility*, is constant.

Since
$$\varepsilon_r = 1 + \chi_e \qquad (3.37)$$

the permittivity is also constant. The capacitance can now be written

$$C = \frac{\varepsilon_r \varepsilon_0 S}{d} \qquad (3.38)$$

The form of this expression explains the reason for the curious choice of symbol ε_r. Many writers call ε_r the *relative permittivity* and the product

$$\varepsilon = \varepsilon_0 \varepsilon_r \qquad (3.39)$$

the *absolute permittivity*. It should be noted explicitly that ε_r is the ratio of two measurements of capacitance, while ε_0 depends on systems of units and cannot be measured independently. Thus ε_r is a pure number, whereas ε_0 is Farad/metre.

The combination of ε_0 and ε_r leads many writers to the erroneous conclusion that the inverse square law in a medium of permittivity ε_r should be written

$$F = \frac{Q_1 Q_2}{4\pi\varepsilon_r \varepsilon_0 r^2} \qquad (3.40)$$

This gives the impression that the medium affects the inverse square law in a rather mysterious manner. It is clearer to explain the actions of the material in terms of the additional surface charges. Once these charges are taken into consideration the material has no further effect. In any calculation we must be careful to take account of all the charges. Once this has been done the calculation proceeds as for any assembly of charges in free space. The forces of the inverse square law cannot be screened off, they act through thick and thin.

One further word needs to be said about E and D in an insulating material. In free space we have written $D = \varepsilon_0 E$, but in a dielectric material it is more convenient to write $D = \epsilon_0\epsilon_r E$. Consideration of Fig. 3.15 shows that with this definition D ignores the induced charges q and depends only on Q. Thus the electric flux of which D is the density takes no account of the insulation, but starts and finishes on a metallic charged surface. D describes the *partial* field due to Q and E describes the *total* field due to Q and q. Thus in a capacitor D depends on the charge on the metal plates and E depends on the voltage. Thus the capacitance depends on the ratio of D to E, and this is the reason why D and E have been defined in this manner.

The calculation of capacitance therefore proceeds in the following manner: Assume that there is a charge on the plates of the capacitor. By the use of Gauss's theorem determine the electric flux and the flux density. Allow for the induced charge on the surface of the dielectric by dividing the flux density by the relative permittivity. Divide by the primary electric constant to find the electric field strength in the dielectric. Hence find the potential difference between the plates. Divide the charge on the plates by the potential difference to find the capacitance. The reader should test his understanding of this process by working the exercises of the end of this chapter.

In our discussion we have assumed ideal insulating materials in which there are no free charges. Actual materials always possess some conductivity and the insulating material allows the charge from the plates to leak away. As long as the time of charge and discharge in a particular application is much shorter than the leakage time, this does not matter greatly. Table 1 gives some useful information about insulating (or dielectric) materials used in capacitors. Figure 3.16 shows three types of capacitor in common use. Both the table and the figure are taken from *Basic Electric Circuits* by A. M. P. Brookes, a companion volume in this series.

TABLE 1

Dielectric	Permittivity (Relative)	Time to lose half of charge	Stability %	Manufacturing accuracy %	Power factor
Vacuum	1	Days	—	—	—
Air	1·0006	Days	± 0·01	± 0·01	0·00001
Polystyrene	2·6	Days	± 0·5	± 20	0·0005
Waxed paper	5	Hour	± 5	± 20	0·01
Mica	6·5	Hour	± 1	± 10	0·002
Ceramic	100	Hour	± 1	± 10	0·001
High permittivity Ceramic	1000	Minute	± 20	± 20	0·01
Electrolyte	—	Second	± 10	± 20	0·05

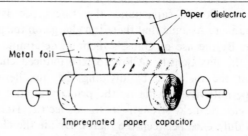

Paper dielectric

Metal foil

Impregnated paper capacitor

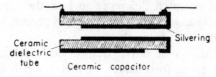

Ceramic dielectric tube

Silvering

Ceramic capacitor

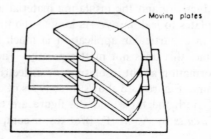

Moving plates

Variable air spaced capacitor

FIG. 3.16 *Typical capacitors*

3.8. CAPACITORS IN PARALLEL AND IN SERIES

Figure 3.17 shows a number of capacitors connected in parallel to the terminals of a battery of voltage V. The total charge supplied by the battery to the capacitors is

$$Q = Q_1 + Q_2 + Q_3 + \ldots \tag{3.41}$$

also
$$V = \frac{Q_1}{C_1} = \frac{Q_2}{C_2} = \frac{Q_3}{C_3} = \ldots \tag{3.42}$$

Hence
$$Q = V(C_1 + C_2 + C_3 + \ldots) \tag{3.43}$$

and the total capacitance is given by

$$C = \frac{Q}{V} = C_1 + C_2 + C_3 + \ldots \tag{3.44}$$

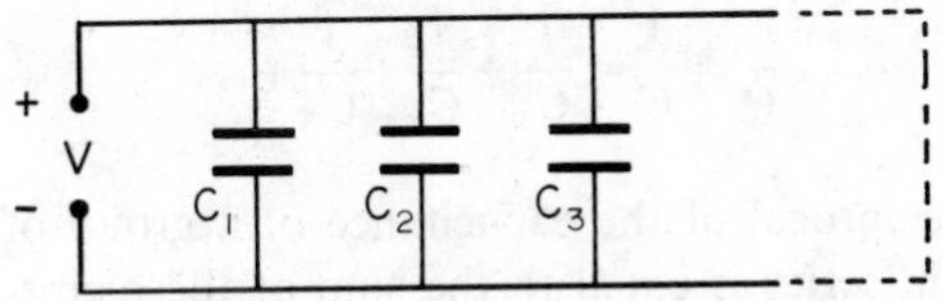

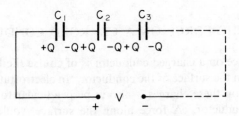

FIG. 3.17   *Capacitors in parallel*

Thus the capacitance of a group of capacitors connected in parallel is the sum of the capacitances.

FIG. 3.18   *Capacitors in series*

Figure 3.18 shows a number of capacitors connected in series to the terminals of a battery of voltage V. Consider the charging process. A charge $+Q$ is placed on the left plate of capacitor C_1. This causes a charge $-Q$ to appear on the right plate of the same capacitor. This in turn produces a charge $+Q$ on the left plate of the second capacitor, since this plate is attached by a conductor to the right plate of the first capacitor. The process is repeated until every capacitor has the same charge.

So
$$Q = Q_1 = Q_2 = Q_3 = \dots \tag{3.45}$$

and
$$V = V_1 + V_2 + V_3 + \dots \tag{3.46}$$

Hence
$$V = \frac{Q_1}{C_1} + \frac{Q_2}{C_2} + \frac{Q_3}{C_3} + \dots$$

$$= Q\left(\frac{1}{C_1} + \frac{1}{C_2} + \frac{1}{C_3} + \dots\right) \tag{3.47}$$

$$\frac{V}{Q} = \frac{1}{C} = \frac{1}{C_1} + \frac{1}{C_2} + \frac{1}{C_3} + \dots \tag{3.48}$$

Thus the reciprocal of the capacitance of a group of capacitors connected in series is equal to the sum of the reciprocals of the individual capacitances.

It follows that capacitance is increased by connecting capacitors in parallel and capacitance is reduced by connecting capacitors in series.

3.9. THE FORCE ON A CHARGED CONDUCTOR

The force on a charged conductor is of course the force on the charges on the surface of the conductor. In electrostatic problems the force on these charges is always perpendicular to the surface of the conductor. A force along the surface would move the charges, but this is not permissible since electrostatics deals with the situation that exists after the motion has ceased. The electrostatic force thus tries to drag the charges out of the surface and, since this is impossible under normal circumstances, the force is transmitted by the surface charges to the whole mass of the body.

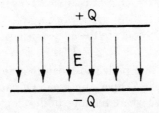

+ Q

E

− Q

FIG. 3.19   *Force on the plates of a parallel-plate capacitor*

Consider the force on one of the plates of the parallel-plate capacitor of Fig. 3.19. Since the field is uniform and since E is Force/Charge, it would seem plausible to write

$$F = EQ$$

This, however, is wrong, because the field varies across the charge from E just outside the plate to zero inside the plate. The average force is therefore

$$F = \tfrac{1}{2}EQ \qquad (3.49)$$

Since

$$Q = D \times S \qquad (3.50)$$

where S is the area of the plate, we can write

$$F = \tfrac{1}{2}E \,.\, DS \qquad (3.51)$$

or

$$f = \tfrac{1}{2}ED \qquad (3.52)$$

where f is the force per unit area.

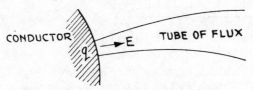

CONDUCTOR   E   TUBE OF FLUX

q

FIG. 3.20   *Force on a conducting surface*

The expression of eqn. (3.52) applies to any conducting surface, as can be seen by reference to Fig. 3.20. Let q be the charge/area on the surface and E the electric field strength just outside the conductor.

Then
$$f = \tfrac{1}{2}Eq \qquad\qquad (3.53)$$

and
$$D = q \qquad\qquad (3.54)$$

whence as before
$$f = \tfrac{1}{2}ED \qquad\qquad (3.52)$$

This result can be interpreted to mean that along a tube of flux there is a tension of $\tfrac{1}{2}ED$ per unit area. Of course this tension is only felt at a surface where the tube of flux comes to an end. The total force on a conductor must be obtained by the vector addition of the forces from all the tubes of flux.

3.10. THE METHOD OF CURVILINEAR SQUARES

In our discussion of capacitors we have so far assumed that the field between the plates is uniform. This is, however, not true at the edges and Fig. 3.21 shows an accurate plot of the field near

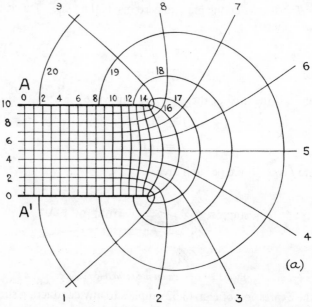

FIG. 3.21 *Field near the edge of a parallel-plate capacitor*

the edge of a parallel-plate capacitor. The field shown is two-dimensional, i.e. the capacitor plates are long in the plane perpendicular to the plane of the paper. The picture looks very complicated, but with a little practice one can make free-hand sketches which are quite accurate. The secret of the process is that all the little areas between adjacent flux lines and equipotentials are almost square. The method is called the method of curvilinear squares and arises from the following considerations.

Refer to Fig. 3.22 and take unit depth into the paper. The flux contained in the tube shown is

$$\delta\Psi = \Psi_2 - \Psi_1 = D\,\delta y \tag{3.55}$$

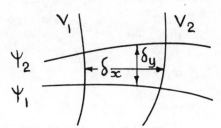

FIG. 3.22   *Flux and potential lines*

and the potential drop $\delta V = V_1 - V_2 = E\,\delta x$ \hfill (3.56)

Hence $$\frac{\delta\Psi}{\delta V} = \frac{D}{E}\frac{\delta y}{\delta x} = \varepsilon_0\frac{\delta y}{\delta x} \tag{3.57}$$

If now we choose the steps of potential difference and of flux,

so that $$\delta\Psi = \varepsilon_0\,\delta V \tag{3.58}$$

we have $$\delta y = \delta x \tag{3.59}$$

Thus with the arbitrary choice of steps in eqn. (3.58) we have achieved a spacing which makes the plot into *curvilinear squares*. In plotting such a picture as Fig. 3.21 one sketches the equipotentials and then tries to make curvilinear squares with the flux lines. At first this may not be possible and both potential lines and flux lines must be adjusted until every part of the plot is in curvilinear squares.*

One use of the resultant plot is to obtain an accurate value of the capacitance. Suppose we wish to know the capacitance per unit length of that part of the capacitor to the right of line AA' in Fig. 3.21.

$$C = \frac{Q}{V}$$

$$= \frac{\Psi}{V} \qquad (3.60)$$

There are 20 tubes of flux and 10 steps of potential.

Hence $\qquad C = \dfrac{20\delta\Psi}{10\delta V}$

$$= 2\varepsilon_0 \frac{\delta y}{\delta x}$$

$$= 2\varepsilon_0$$

$$= 17 \cdot 7 \times 10^{-12} \text{ Farad/unit length} \qquad (3.61)$$

Often it is convenient to obtain flux plots such as in Fig. 3.21 by experiment. This can be done by constructing suitable models and by measuring the equipotentials in an electrolytic tank or on a sheet of conducting paper. In such cases one is really finding the field associated with an electric current, but we shall show in the next chapter that this also is an electrostatic field.

SUMMARY

Most of this chapter has been concerned with the electric field. We have discovered that a field map showing lines of electric field strength can be used to obtain information about the strength of the charges causing the field.

* Note that this method is only applicable where there is no variation of the field in the z direction.

We have proved Gauss's theorem which states that the integral of $\varepsilon_0 E_n$ over a closed surface is equal to the charge Q within that surface.

Gauss's theorem has been used to calculate the field of certain distributions of charge.

We have invented the terms electric flux density and electric flux and have shown that flux may be conveniently split into *tubes of flux*.

A consideration of the storage of energy in the electric field has led us to describe capacitors and we have shown how insulating materials may be used to increase capacitance.

We have shown how to calculate forces on charged conductors and finally have introduced the method of curvilinear squares as an accurate way of plotting two-dimensional electric fields.

NEW TERMS USED IN THIS CHAPTER

Term	Symbol	Unit (Abbreviation)	Definition
Flux Density	D	coulomb/m² C/m²	$D = \varepsilon_0\varepsilon_r E$
Flux	Ψ	coulomb C	$\Psi = \iint_S D_n \, dS$
Capacitance	C	farad F	$C = \dfrac{Q}{V} = \dfrac{\Psi}{V}$
Permittivity (relative)	ε_r	Pure number	$\varepsilon_r = \dfrac{C}{C_0}$
Susceptibility	χ_e	Pure number	$\chi_e = \varepsilon_r - 1$
Primary Electric Constant	ε_0	farad/m	$F = \dfrac{Q_1 Q_2}{4\pi\varepsilon_0 r^2}$

Exercises

3.1. A coaxial cable has an inner conductor of diameter 1 mm and an outer conductor of internal diameter 9 mm. The insulation between conductors is polythene of permittivity $\varepsilon_r = 2\cdot3$. What is the capacitance of 10 km of the cable?
(*Ans.* $0\cdot582\,\mu$F).

3.2.　Derive an expression for the maximum voltage allowable between long coaxial conducting cylinders of radii R_1, R_2 ($R_2 > R_1$), if the field strength in the insulation between them is not to exceed E_m.

(*Ans.* $E_m R_1 \ln R_2/R_1$).

3.3.　Determine from first principles the capacitance per metre of a concentric cable whose inner conductor has radius r_1 and whose outer conductor has internal radius r_3, when the intervening dielectric is distributed as in Fig. 3.23, ε_1 and ε_2 being the (relative) permittivities.

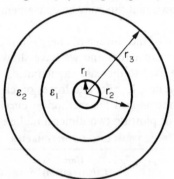

FIG. 3.23　*Relating to exercise* 3.3

What is the energy stored per metre length when $r_1 = 17$ mm, $r_2 = 22$ mm, $r_3 = 30$ mm, $\varepsilon_1 = 4$, $\varepsilon_2 = 2$ and the p.d. between inner and outer conductor is 33 kV?

$$\left(Ans. \quad \left[\frac{2\pi\varepsilon_0}{\dfrac{1}{\varepsilon_1} \ln \dfrac{r_2}{r_1} + \dfrac{1}{\varepsilon_2} \ln \dfrac{r_3}{r_2}} \right], \; 0.14 \text{ joule} \right)$$

3.4.　A capacitor consists of a pair of large parallel metal plates 3·0 mm apart, between which lies a parallel sheet of mica 2·5 mm thick; air fills the rest of the space between the plates. If the mica be taken to have a permittivity of 5, find the density of charge on the plates, and the voltage gradient in the air, when the capacitor is charged to a potential difference of 2,000 V.

If the mica is replaced by a pair of thin parallel metal sheets separated by 2·5 mm of air, find the density of charge needed on these sheets in order that the charge on the outer plates shall be the same as when the mica was in place, the p.d. of the capacitor being kept constant at 2,000 V. (*Ans.* 17·6 × 10⁻⁶ Coulomb/m², 2·0 MV/m, 14·2 × 10⁻⁶ Coulomb/m²).

3.5.　A capacitor consists of two plane parallel plates each 25 cm² in area and 2 mm apart. A sheet of dielectric 1 mm thick and relative permittivity 2·5 is placed between the plates. Neglecting edge effects, estimate the capacitance.

What would be the capacitance of a capacitor which is geometrically

similar but reduced in scale by a factor of ten?
(*Ans.* 15·8 pF, 1·58 pF)

3.6. The permittivity of an insulating circular disc of 7cm diameter and 1cm thickness is checked by placing it between two plates (diameter $\gg$ 7cm), and noting the difference in their capacitance with and without the disc, the plates being 1 cm apart. This difference is found to be 240 pF. If, when the disc is present, there is an average air-gap between each plate and the disc surface of 0·01 mm, find the error in estimating the permittivity on the assumption that the plates are in contact with the disc over the whole of the surface.
(*Ans.* 11·8 pF low)

3.7. The space between the "plates" of a spherical capacitor has inner radius a and outer radius b and is occupied by a material of permittivity ε_r. Calculate the capacitance.

$$\left(Ans. \ 4\pi\varepsilon_0\,\varepsilon_r \middle/ \left[\frac{1}{a} - \frac{1}{b}\right]\right)$$

3.8. Explain what is meant by a tube of flux.

In a certain electrostatic field a conducting sheet is placed along an equipotential surface. The charges within the volume enclosed by the conductor are then removed. What charge must be placed on the conductor so that the field outside the surface is unchanged? Show carefully why there will be no electric field in the volume enclosed by the conductor.

3.9. Show that the potential energy stored in a capacitor can be expressed as the volume integral of $\frac{1}{2}ED$ throughout the volume occupied by the electric field. (*Hint:* Draw tubes of flux and equipotential surfaces and consider each part of the field as a small capacitor.)

Check the validity of the method by finding the energy stored in a spherical capacitor.

3.10. Develop an expression for the force on the plates of a parallel-plate air capacitor in terms of the capacitance, the p.d. and the distance between the plates.

Calculate the force if the plates are 2000 mm² and 3 mm apart and if the p.d. is 500 V.

$$\left(Ans. \ \frac{1}{2}\frac{CV^2}{d}, \ 2\cdot46 \times 10^{-4} \ \text{newtons}\right)$$

3.11. Figure 3.21 shows the flux at the edge of a pair of long parallel plates. To the left of the region shown the field within the plates is practically uniform, and outside them it is negligible. The width of the plates is large compared with the distance between them. Estimate the error involved in finding the capacitance between the plates per unit run, if edge effects are neglected.
(*Ans.* $-0\cdot5\varepsilon_0$ Farads)

3.12. An electrostatic voltmeter consists essentially of nine parallel semi-circular plates, alternate plates being connected together. One set of plates is fixed, the other set is free to move; the spacing between adjacent plates is 4 mm, and the plate diameter is 100 mm. The medium between the plates is air.

What is the total capacitance when the moving plates have swung through one radian from a position of approximately zero capacitance?

What should be the stiffness of the suspension to give 120° rotation to the moving plates for a p.d. between plates of 1,000 V? Neglect edge effects.

(*Ans.* 22·1 pF, 5·28 × 10^{-6} Nm/rad)

(*Hint:* For the second part use the principle of virtual work, i.e. displace the plates slightly, keeping the charges constant.)

CHAPTER 4

Steady Electric Currents

4.1. TYPES OF ELECTRIC CURRENT

In the last two chapters we have been considering the problems of electrostatics. In all such problems the electric charges are at rest and interaction between the charges is due to their relative position. The remainder of this book will be taken up with the more interesting, but also more complicated, effects associated with the motion of charges, that is with *electric currents*. A current of 1 ampere is defined as the flow of 1 coulomb per second. This chapter is concerned with charges in uniform motion. There are various types of electric current and it is helpful to distinguish between them.

4.1.1. *Conduction current*

In a metallic conductor an electric field will set up an electric current by moving the free electrons through the stationary atomic lattice. The positive direction of current flow is taken as the direction of the electric field strength. This direction is from a positive to a negative charge and is therefore the direction in which a positive charge would move. This means that the actual motion of the negatively charged electrons is in the opposite direction to that of the supposed current.

In semiconductors one has to consider the motion of electrons and holes, i.e. both negative and positive charge carriers.*

4.1.2. *Convection current*

Secondly there is the convection current which does not take place in a solid conductor but in space. A possible example of a convection current would be the motion of a charged body carrying its charges with it. A more usual case is that of vacuum tubes in which electrons are evaporated from a hot cathode and propelled by the electric field towards the anode. In such devices the convection current consists again entirely of electrons and the flow of the conventional current is in the opposite direction from anode to cathode. In gas discharge tubes, on the other hand, there is a motion of positive ions towards the cathode as well as of electrons towards the anode. The total current is then the sum of the charges per second transferred by the ions and the electrons. In liquids also the charge may be transferred by positive and negative ions.

The division into conduction and convection currents is somewhat arbitrary and it is quite usual to speak of current conduction in gases and liquids. Nevertheless, because of the outstanding importance of metallic conductors in electrical engineering, it is useful to treat them as a class by themselves and to regard the other processes in gases and liquids, and in a vacuum as the convection of charge.

4.1.3. *Polarization current*

The third type of current is that which occurs in insulators. Here the charges are not free to move through the material, but the straining of the atoms and the alignment of polar molecules result, as was discussed in section 3.7, in a transfer of electric

* See Harris and Robson, *The Physical Basis of Electronics*, a companion volume in this series.

charge from one face of the insulator to the other. To an observer outside the material this effect is indistinguishable from a motion of free charge. We, therefore, call this a polarization current.

4.2. OHM'S LAW

We come now to a more detailed discussion of conduction currents. Consider a charged capacitor as in Fig. 4.1 and suppose

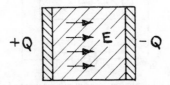

Fig. 4.1 *A charged capacitor*

that the insulating material between the plates has free charges in it. It is in fact a "leaky" dielectric. Under the action of the field E the free charges will move and the capacitor will be discharged (Fig. 4.2). Suppose now that a device is connected to the

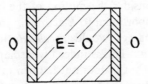

Fig. 4.2 *The capacitor has been discharged*

capacitor plates such that the charge on the plates is always renewed at the same rate as it leaks away. The capacitor now remains charged and a constant current flows. A device that supplies charge in the way described is called a *generator* (Fig. 4.3).

Since the charge on the capacitor plates remains constant, the p.d. between the plates will also remain unchanged. The generator

EFE—F

does not have to be a reservoir of charge but can be made in the form of a pump which takes the charges from one plate of the capacitor and raises them through the necessary p.d. in order to restore them to the other plate. If for the sake of argument we think of the positive charges as moving, then a positive charge

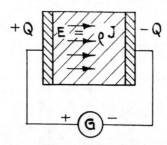

Fig. 4.3   *Leaky capacitor continually recharged by generator*

moves through the slab under the action of the field E. It loses energy and arrives at the negative plate at a lower potential. This energy is restored by the generator which returns the charges to the positive plate. The fact that the negative charges move and the positive charges are fixed leaves the argument unimpaired.

We must now consider the motion of the charges in the slab. There is a force proportional to E acting on them and one would expect that the charges would *accelerate* in accordance with Newton's second law. This is indeed what happens in a diode or other vacuum tube, but it does not happen in a conducting slab. A rather surprising result is observed. A constant electric field strength E causes the charges to move with *constant velocity*. The force is thus proportional to the velocity of charge and hence to the current. This experimental fact is known as Ohm's law and can be written

$$E = \rho J \qquad (4.1)$$

where J is the current density at the point at which E acts and ρ is the constant of proportionality and is called the *resistivity*. Its unit is the volt-metre/ampere or ohm-metre, since volt/ampere is called the ohm. Sometimes it is more convenient to describe the same result by the equation

$$\sigma E = J \qquad (4.2)$$

where $\sigma = 1/\rho$ is called the *conductivity*.

A clue to the behaviour of conductors is found in the similar result in fluid mechanics. In studying the seepage of water through soil one finds that the force acting on the water is proportional to the velocity and not to the acceleration of the water. Current flow is a sort of seepage. The electrons are in a state of more or less random motion and collide with one another and with the positive charges. What we observe as a velocity is really an average drift velocity. Individual electrons are accelerated by the field between collisions, but they lose their extra kinetic energy at each collision and thus their mean velocity is constant.

The energy that is lost in the collisions heats the conductor. The rate of energy loss is given by the p.d. multiplied by the charge per second ($V \times I$). The ratio of p.d. to current is called the resistance R and its unit is the ohm. If the conductor has uniform cross-sectional area S the current density will be uniform and by eqn. (4.1) E will be constant along the conductor.

Hence $\qquad\qquad\qquad V = El \qquad\qquad\qquad (4.3)$

and $\qquad\qquad\qquad I = JS \qquad\qquad\qquad (4.4)$

Then $\qquad\qquad\qquad R = \dfrac{V}{I} = \dfrac{El}{JS}$

$$= \rho \frac{l}{S} \qquad (4.5)$$

Equation (4.5) is a special case of the general Ohm's law [eqn. (4.1)], but it is the usual form in which Ohm's law is expressed. Ohm's law is experimental, a theoretical proof of it would involve extremely complicated investigations into the micro-structure of conductors. The law states that the resistance R, or the resistivity ρ, are independent of the magnitude of the current and the electric field strength. This is an extraordinarily useful piece of information and simplifies our calculations greatly. However, R depends on temperature, because in general an increased temperature would result in an increased number of collisions and hence in an increased energy loss. The ohmic resistance also depends on the metal used. For any particular metal the temperature dependence may be written:

$$R_T = R_1 \left[1 + \alpha_1 (T - T_1) \right] \tag{4.6}$$

where R_T is the resistance at temperature T, R_1 is the resistance at temperature T_1, and α_1 is the *temperature coefficient of resistance* for the material appropriate to T_1. Table 4.1 gives temperature coefficients of resistance for various metallic substances. These coefficients are referred to a temperature $T_1 = 0°C$ and to the Celsius (Centigrade) scale of temperature.

TABLE 4.1.*

TEMPERATURE COEFFICIENTS OF RESISTANCE

Manganin	0·000002
Eureka	− 0·00007 to + 0·00004
Nichrome	0·0001
Brass	0·001
Gold	0·0034
Platinum	0·0037
Silver	0·0038
Copper	0·0043
Lead	0·0043
Aluminium	0·0043
Tungsten	0·0045
Iron	0·0055
Nickel	0·0059

* Tables 4.1 and 4.2 are taken from *Basic Electric Circuits* by A. M. P. Brookes, a companion volume in this series.

Table 4.2 gives values of the resistivities of certain metals.

TABLE 4.2.*

RESISTIVITIES OF METALS AT 20°C

	ohm-metre $\times 10^{-8}$
Silver	1·62
Copper	1·76
Gold	2·40
Aluminium	2·83
Tungsten	5·48
Nickel	7·24
Brass	8
Iron	9·4
Platinum	10
Lead	20
Manganin	45
Eureka	49
Nichrome	108

Ohm's law [eqn. (4.1)] shows that there has to be an electric field at every point at which there is a current flow. In the special case of Fig. 4.3 the field arises from the charges on the plates and on the surface of the partially conducting block. In this case we can write

$$Q = \varepsilon_0 E \times \text{Area} \qquad (4.7)$$

Also $$E \times \text{Area} = \rho J \times \text{Area}$$

$$= \rho I \qquad (4.8)$$

Therefore $$Q = \varepsilon_0 \rho I \qquad (4.9)$$

Thus, if the supposed leaky dielectric is in fact a metallic conductor such as copper,

$$Q = 8·854 \times 10^{-12} \times 1·76 \times 10^{-8} I \qquad (4.10)$$

* See note p. 68.

Thus for a current of 100 amperes we should need a surface charge of

$$Q = 1 \cdot 56 \times 10^{-17} \text{ coulombs} \qquad (4.11)$$

The surface charge is thus extremely small in the case of a metal. If on the other hand the leaky dielectric is a material such as a ceramic of resistivity 10^{12} ohm-metres, then for a current of 100 amperes the surface charge would have to be

$$Q = 885 \cdot 4 \text{ coulombs} \qquad (4.11')$$

which is an impossibly large surface charge and would cause electrical breakdown.

These two examples show why we can make the sharp division into conducting and insulating materials and why the same geometrical arrangement can be regarded as a capacitor in one case and a conductor (or resistor) in another case. In both cases there are surface charges and conduction current, but the ratio between the two quantities is vastly different.

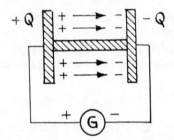

FIG. 4.4   *Conductor between two charged plates*

Consider now a thinner conductor between the same plates (Fig. 4.4) and let the cross-section of the conductor be small compared with the cross-section of the plates. In this case the field E causing the current to flow is still due to the charges on the plates only. Now let the conductor be bent as in Fig. 4.5. Ohm's law requires the field strength E to be along the conductor, but such a field cannot be produced by the charges on the plates. The field has to be distorted and the only way in which this can come about is by means of an additional charge distribution. There can be no charges within the volume of the conductor and the additional charges must therefore lie on the surface.

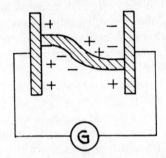

FIG. 4.5 *A curved conductor between plates*

We can generalize the result and say that conductors always have a surface charge distribution which gives the correct field to produce current flow. When the generator is first connected a transient current flows to distribute the surface charge, only then can the steady conduction current establish itself. In other words, every conductor has capacitance as well as resistance when there is a current flow within it. This becomes important at very high frequencies because the surface charge contributes to a so-called "stray" capacitance. Even in low-frequency problems or problems of steady current-flow it is essential to remember the surface charges in order to achieve a clear understanding of the mechanics of the problem. The surface charges enable the conductor to carry its field with it, and this has the very useful result that wires can be bent into any desired shape without altering the resistance.

Because the field is due to stationary surface charges, it is an electrostatic field. This explains the use of conducting models, such as in an electrolytic tank, in the examination of the electric fields of apparatus in which no current flows. This use has already been mentioned in our discussion of the method of curvilinear squares in Section 3.10.

4.3 DRIFT-VELOCITY OF ELECTRONS IN A CONDUCTOR

A knowledge of the electric current does not by itself give any information about the mean velocity of the charges. Since electrical effects can be transmitted with the speed of light 3×10^8 m/sec, one might expect that the drift-velocity would be very high. This, however, is not so, because the free charge in conductors is enormous. Thus in copper, where there is one free electron per atom, the available charge is $8 \cdot 1 \times 10^{28}$ electrons per cubic metre. This is a charge of $1 \cdot 3 \times 10^{10}$ coulomb. Thus for a current density of $1 \cdot 5 \times 10^6$ amperes/m^2, which is commonly used in wires, the velocity is given by

$$v = \frac{A/m^2}{Q/m^3} = \frac{1 \cdot 5 \times 10^6}{1 \cdot 3 \times 10^{10}}$$

$$= 1 \cdot 15 \times 10^{-4} \, \text{m/sec} \tag{4.12}$$

Thus the velocity is approximately 1/10 mm/sec. It can, therefore, be seen that for the propagation of an electric signal along a wire there is no need for the motion of an electron from one end of the wire to the other. All that is needed is that the forces on the electrons should be transmitted. The other point to remember is that the velocity of eqn. (4.12) is not the velocity of all the electrons, but is an average drift-velocity. The electrons themselves will have velocities of different magnitudes in all directions.

4.4. ELECTROMOTIVE FORCE

We found in Section 4.2 that the electric field which drives current through a conductor is the electrostatic field of charges on the surface of the conductor. This faces us with an apparent paradox. Electrostatic fields are conservative, energy cannot be continuously abstracted from such fields. But in the conductor, electrical energy is continuously being converted into heat at the rate of

$$w = EJ = \rho J^2 \, \text{watt/m}^3 \tag{4.13}$$

or
$$W = VI = RI^2 \, \text{watts} \tag{4.14}$$

This energy must come from somewhere and we now remember that the isolated conductor is not sufficient to pass a continuous current, there must also be a generator. Inside the generator there must be non-conservative forces which are able to transfer the charges from a lower to a higher potential against the electrostatic forces. The charges are moved "up the hill" because the non-conservative forces in the generator are stronger than the conservative forces in the generator. Outside the generator there are only conservative forces.

How do the non-conservative forces arise? In large generators they are produced by driving conductors through magnetic fields. We shall discuss this in Chapter 6. In such generators, mechanical energy is converted into electrical energy. Another common method is used in the ordinary chemical battery, in which chemical energy is converted into electrical energy. Both these processes have drawbacks and much research is being devoted at the present time to the development of a process in which heat energy might be converted into electrical energy without the intervention of steam boilers and turbines.

To the outside world the electrical effect of a generator is its provision of a potential difference at its terminals. It acts like a capacitor which cannot be discharged, because the charges on the plates are continuously replenished by some internal action. In order to distinguish this potential difference it is given the name *Electromotive Force* and is generally written e.m.f. Its unit is, of course, the volt.

The term electromotive force is not a very happy choice. The "electromotive" part is admirable, but the term "force" is misleading. As in the case of p.d. the dimensions of e.m.f. are energy/charge and not force. But the term is so well established

that it seems unlikely that it will be changed, and so generations of
students have to be taught that this "force" is not a force.

Figure 4.6 shows an idealized circuit in which a generator is
circulating current through a resistor. Outside the generator the

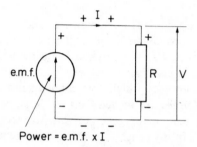

Power = e.m.f. x I

FIG. 4.6   *Ideal generator supplying power to resistance*

field is determined by electrostatic forces due to surface charges,
which are roughly as indicated. For the resistor we have

$$V = IR \qquad (4.15)$$

and
$$\text{Power} = VI = I^2R \qquad (4.16)$$

For the generator

$$\text{e.m.f.} = V \qquad (4.17)$$

and
$$\text{Power} = \text{e.m.f.}\, I \qquad (4.18)$$

Thus the electrical energy from the generator is converted to heat
in the resistor at the same rate at which it is being generated. The
arrow drawn in the circle showing the generator serves as a
reminder that the energy has to be supplied to the generator in the
form of chemical, mechanical, heat or some other kind of energy.

The generator in Fig. 4.6 is idealized to have no losses. It is
thus purely a source of electromotive force. Actual generators
have losses which convert some of the electrical energy into heat.
A more accurate way of describing generators is therefore by

means of Fig. 4.7 where the losses are represented by an equivalent resistor. This is the usual way of making allowance for loss and it does not imply that there is inside the generator a metallic conductor of resistance r. There may be such a conductor or there

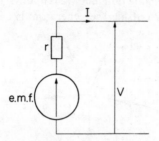

FIG. 4.7   *Generator with internal loss*

may not. When there is no current flowing, the generator is said to be open-circuited and the p.d. V will then be equal to the e.m.f. This is the way of measuring the e.m.f. When a current I flows through the generator, the p.d. will drop to the value

$$V = \text{e.m.f.} - Ir \tag{4.19}$$

Consideration of Fig. 4.8 shows that we can make the general statement that: "The sum of the electromotive forces around any

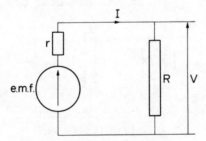

FIG. 4.8   *Kirchhoff's second law*

closed circuit is equal to the sum of the IR drops around the circuit." This statement is known as Kirchhoff's second law.

Kirchhoff's first law states that the total current arriving at any point in an electric network must be zero. Thus in Fig. 4.9 $I_1 + I_3 + I_5 = I_2 + I_4$.

Kirchhoff's first law applies only if there is no means of storing the charge at the junction and Kirchhoff's second law applies only if the energy outside the generators is being entirely lost as

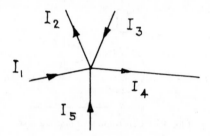

FIG. 4.9   *Kirchhoff's first law*

heat energy. Both laws, therefore, apply to the case of steady current flow discussed in this chapter, but have to be modified if the currents are varying.

The representation of a generator as an e.m.f. in series with a resistance is very common and useful. Often, but not invariably, the losses in a generator are small and it can be regarded as a

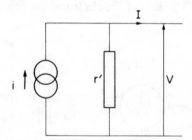

FIG. 4.10   *A current source*

source of constant voltage. This is the way in which the electric supply system is arranged all over the world. The customers are supplied with current at practically constant voltage whether the current taken is large or small. There are, however, special applications where the current is kept substantially constant and the e.m.f. is altered to make this possible. In such applications a suitable representation of the generator is that of Fig. 4.10, in which the generator is shown as a source of constant current in parallel with a resistance to account for its internal loss. Since the circuits of Figs. 4.7 and 4.10 are to be equivalent, we can write

$$V = \text{e.m.f.} - Ir$$

$$= (i - I)r' \tag{4.20}$$

On open circuit $I = 0$ and therefore

$$\text{e.m.f.} = ir' \tag{4.21}$$

hence from eqn. (4.20)

$$ir' - Ir = ir' - Ir' \tag{4.22}$$

and

$$r = r' \tag{4.23}$$

Thus the equivalent resistances are equal in both representations and the current of the current generator is equal to the e.m.f./internal resistance.

In books on electric circuits Fig. 4.7 is often called the Thèvenin equivalent voltage generator and Fig. 4.10 is called the Norton equivalent current generator.

SUMMARY

After listing various types of electric current we have devoted most of this chapter to an examination of conduction in metallic conductors.

We have discussed Ohm's law $E = \rho J$ and found that it implies that conduction is similar to the seepage of water through soil. We have shown that an electrostatic field is associated with steady currents and that this field is in general due to charges on the surface of conductors.

Consideration of the drift-velocity of electrons through conductors has shown that this velocity is surprisingly small.

Finally we have discussed the energy balance in electric circuits and have shown that electromotive force is necessary to circulate steady currents.

NEW TERMS USED IN THIS CHAPTER

Term	Symbol	Unit (Abbreviation)	Definition
Current	I, i	ampere A	Flow of Charge (Coulomb/sec)
Current Density	J	ampere/metre² A/m²	$\iint_S J_n \, dS = I$
Resistivity	ρ	ohm-metre Ωm	$E = \rho J$
Conductivity	σ	siemen/metre s/m	$\sigma = \dfrac{1}{\rho}$
Resistance	R, r	ohm Ω	$R = V/I$
Temperature Coefficient of Resistance	α	—	$R_T = R_1[1 + \alpha_1(T - T_1)]$
Power	W	watt W	$W = VI$
Electromotive Force	e.m.f.	volt V	e.m.f. $= V$ for a generator on open circuit

Exercises

4.1. What is meant by current density? Show that this quantity is a vector and explain how you would find the current through a curved surface of arbitrary shape, if the current density at the surface were given.

4.2. Give a general account of Ohm's law and state what is meant by resistance.

4.3. A variable resistor is connected to the terminals of a generator of constant e.m.f. and internal resistance. Prove that the power supplied to the resistor is greatest when its resistance is adjusted to be equal to the internal resistance of the generator.

4.4. In Section 3.8, rules were developed for finding the equivalent capacitance of capacitors connected in parallel and in series. Develop similar rules for the equivalent resistance of resistors connected in parallel and in series.

4.5. State Kirchhoff's two laws. Show how the second law can be deduced from the knowledge that an electrostatic field is conservative.

4.6. Why is it possible to find the electric field between two electrodes in air by means of an experiment in which scale-models of the electrodes are placed in a tank containing a conducting fluid? What corresponds in the electric field to the equipotentials and current flow lines of the experimental model?

4.7. A liquid of resistivity ρ is contained in a cylindrical vessel of length L and diameter D. The top and bottom of the vessel are insulators, but the curved surface is made of conducting material of negligible resistance. Along the axis of the vessel there is placed a cylindrical conducting rod of length L and diameter d $(d < D)$ also of negligible resistance. Obtain an expression for the voltage v required to produce a total current i from the rod through the liquid to the container wall. If $\rho = 0.5$ Ωm, $L = 10$ m, $D = 3$ m, $d = 0.5$ m, find the voltage which must be applied if 10 kW of heating power is required for the liquid.
    (*Ans.* 11.9 V)

4.8. The main generators of a ship are being used to charge her batteries, the open-circuit p.d. of which is 199.0 V. One generator has an e.m.f. of 202.5 V and equivalent series resistance of 0.0035 Ω; the other has an e.m.f. of 200.75 V and a series resistance of 0.00525 Ω. The generators are connected in parallel to the batteries; and the internal resistance of the latter and of the connections from the generators totals 0.00175 Ω. What current is supplied?
    (*Ans.* 728 A)

4.9. The dielectric between the conductors of a concentric cable has internal and external diameters of 17 and 27 mm. The permittivity is 3.0 and the resistivity of the dielectric is 15 × 10^{12} Ωm. Calculate the capacitance and resistance between the conductors per metre length of the cable.
    (*Ans.* 360 pF, 1.1 × 10^{12} Ω)

4.10. A cable forming a "ring main" is 500 yd long and contains two conductors each of resistance 0.025 Ω per 100 yd. A constant p.d. of 230 V is maintained between the conductors at the point of supply. Loads taking the following currents are connected to the main, the distances giveñ being measured in one direction round the ring from the point of supply:

Concentrated loads of 120 amperes at 150 yd, and 90 amperes at 300 yd

Uniformly distributed load of $\frac{1}{4}$ ampere/yd over the first 200 yd

Find the p.d. between the conductors at the 120 amperes load and the power lost in the main.

(*Ans.* 219·4 V, 2·51 kW)

4.11. A pair of uniform cables runs between substations at its ends and supplies a given current load which is uniformly distributed along the mains. Show that the loss of power due to resistance is least when the p.d.s at the substations are the same.

CHAPTER 5

The Magnetic Field of Steady Electric Currents

5.1. ELECTRICITY AND MAGNETISM

In Section 1.5 it was pointed out that the study of electricity is essentially the study of the mechanics of electric charges. Such a study naturally falls into the two divisions of mechanics, namely statics and dynamics. Electrostatics was discussed in Chapters 2 and 3 and in the last chapter we began a discussion of electrodynamics by considering the important case of conduction in metals. It is clear that logically the next step would be to consider more general cases of the motion of charges and of the forces associated with such motion. But this is one of the parts of the subject where a little knowledge of history can save a great amount of unnecessary work later on. Instead of making a frontal attack on the problems of electrodynamics we shall do some skirmishing near the edges of the subject. Our method may appear a little circuitous, but it will avoid difficult mathematics and the memorization of a number of very ill-conditioned formulae.* Our development of the subject will be similar to the historical development.

---

* One example is Ampère's formula for the force between two current elements

$$F = \frac{1}{2} \frac{ii' \, \mathrm{d}s \, \mathrm{d}s'}{r^2} \left(2 \cos \varepsilon + 3 \frac{\mathrm{d}r}{\mathrm{d}s} \frac{\mathrm{d}r}{\mathrm{d}s'} \right)$$

where ε is the angle and r the distance between the elements of lengths $\mathrm{d}s$ and $\mathrm{d}s'$ and current i and i'.

81

This is not surprising because the original workers confronted the same difficulties that we are facing and they took the easiest road they could find through those difficulties.

The clue that we shall follow is given in the title of this book. Why is the subject generally known as electromagnetism? What is magnetism and why is it linked with the subject of electricity?

The earliest workers such as Gilbert (in A.D. 1600) treated electricity and magnetism as two separate subjects and magnetism was to them the more interesting of the two. For one thing it was useful in navigation, and for another, one could do some very pretty experiments with natural magnets, whereas static electricity was both useless and feeble in its effects. Remember that there were no batteries or other generators at the time.

It was found that magnetic effects were strongest near the ends of magnets and these ends were called the poles. It was also found that poles always occur in pairs of equal and opposite strength, so that it is impossible to isolate one type of polarity. Gilbert pointed out that the earth is itself a great magnet and has two poles. He also found that like poles repel and unlike ones attract. He did not, however, enunciate a law of interaction between magnets, because he knew that this was complicated and in any case he thought more about magnetic fields than about magnetic pole strength.

It was Michell in 1750 who took an enormous step forward. Instead of dealing with the interaction of magnets (i.e. of dipoles) he had the idea of thinking of the poles individually and he found the amazing result that the interaction of magnets can be explained on the basis of an inverse square law between poles

$$F = \frac{\bar{Q}_1 \bar{Q}_2}{4\pi\mu_0 r^2} \tag{5.1}$$

where $\bar{Q}_1$ and $\bar{Q}_2$ are magnetic pole strengths, μ_0 is a constant depending on the system of units and r is the distance between the poles. This does not of course imply that there are in magnetism free charges as there are in electricity. Magnetic poles are analogous not to free charges, but to the polarization charges which are found in insulators. The smallest magnetic entity is a dipole and not a single pole. If it had not been for the discovery that the inverse square law applied to the ends of the dipole taken one at a time, nobody would have bothered with magnetic poles. But Michell's discovery was such an amazing stroke of good fortune that electrical engineers have used the idea of magnetic *poles* ever since. It is wonderful enough to be able to use the mathematics developed for gravitational problems also for electrostatics. Now we can use exactly the same methods also for magnetostatic problems. All three phenomena are based on an inverse square law of force. Here is a tremendous saving of valuable memory space.

5.2. MAGNETOSTATIC FIELDS

The development of ideas and formulae for magnetic problems can now proceed in the same manner as for electric fields. Instead of using the inverse square law directly it is convenient to proceed in two steps as before. The sources produce a field and the field acts on the poles in which we are interested. We, therefore, define a magnetic field strength H, where

$$H = \frac{\bar{Q}}{4\pi\mu_0 \, r^2} \tag{5.2}$$

which is analogous to E the electric field strength.

Work has to be done in assembling magnetic poles and we define a potential or potential difference, which is the work done in carrying the pole divided by the pole strength. Thus

$$\bar{V} = \int_1^2 -H \, dl \tag{5.3}$$

Thus the magnetic field strength is the gradient or slope of the potential. The negative sign in the expression arises from the fact that work has to be done against the field strength in order to increase the potential.

Whereas H, like E, is a vector quantity, the potential associated with a place is just a number (so many units of energy per unit of pole strength). It is possible to draw a potential map with equipotentials marked on it, and the difference in potential between two places is independent of the path taken between the places. The system is conservative.

It is also convenient to define a magnetic flux density B, where

$$B = \mu_0 H * \qquad (5.4)$$

and a magnetic flux Φ, where

$$\Phi = \iint_S B_n \, dS \qquad (5.5)$$

Over a closed surface S we can apply Gauss's theorem

$$\iint_S B_n \, dS = \sum \bar{Q} \qquad (5.6)$$

where $\sum \bar{Q}$ is the total pole strength inside the volume enclosed by the surface S. Since there are no free poles, eqn. (5.6) simplifies to the statement that

$$\iint_S B_n \, dS = 0 \qquad (5.7)$$

A little care is needed in the application of Gauss's theorem to magnets. In Fig. 5.1 there is no difficulty because the magnet is

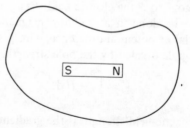

FIG. 5.1 *Magnet surrounded by Gaussian surface*

* This is true in free space. In a magnetic material the flux is stronger and we then write $B = \mu_0 \mu_r H$. See Section 7.4.

entirely surrounded by the surface and eqn. (5.7) is clearly correct, because there is no net pole strength in the volume considered. In Fig. 5.2 it looks as though eqn. (5.7) is incorrect. It all depends what happens at that part of the surface which crosses the magnet.

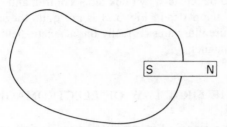

FIG. 5.2 *Magnet projecting through surface*

In order to make eqn. (5.7) apply to all such cases we make the stipulation that the surface shall not pass through the material of the magnet but through free space. The magnet is assumed to be broken as in Fig. 5.3. With this assumption the total magnetic flux through a closed surface is always zero.

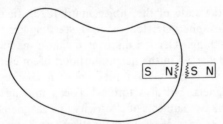

FIG. 5.3 *Magnet broken where surface meets it*

As in electrostatics, so in magnetostatics, it is convenient to divide the flux into tubes. These tubes can be supposed to end on the pole strength on the surface of a magnet. We shall discuss in a later chapter what happens inside a magnet.

Except for the absence of free poles the parallelism between electrostatics and magnetostatics is complete. Consider for instance Fig. 3.21, page 56. This was drawn to illustrate the electric field between the plates of a capicitor. It does equally well as an illustration of the magnetic field between two magnetized surfaces. The electric potential lines are now lines of magnetic potential and the electric flux density is magnetic flux density. The only difference is that in an actual magnetic structure the plates would be replaced by thick blocks of iron and so the field at the back of the plates in Fig. 3.21 is not quite the same. Otherwise the figure illustrates correctly the magnetic field between the faces of a magnet.

5.3. THE FIRST LAW OF ELECTROMAGNETISM

The twin sciences of electricity and magnetism grew up more or less independently of one another. It was known that thunderstorms sometimes interfered with magnetic measurements at sea, but apart from this there seemed to be little connection between the two phenomena. There was of course the mathematical link of the inverse square law but that might have had no physical significance.

This was the state of the subject until 1820. In that year the Danish philosopher Oersted made a startling discovery which altered everything. Oersted discovered that a magnetic compass needle was deflected in the neighbourhood of an electric current. Thus there were not only forces between charges and forces between magnets, but also forces between moving charges and magnets. The two subjects of electricity and magnetism had come together, the science of electromagnetism was born.

Scientists everywhere realized the crucial nature of Oersted's experiment and studied its implications. Foremost amongst them was Ampère, who within three months of hearing the news had developed a complete theory of the interaction of electric currents and magnets and of the action of electric currents upon one another. Ampère carried out a series of brilliant experiments and also developed a mathematical theory. The work earned for him the title of the "Newton of electricity". He found out, amongst other things, that parallel currents attract if they flow in the same direction and repel if they flow in opposite directions. His discoveries can be summarized in the first law of electromagnetism: "The magnetic field of a small current loop is identical with the magnetic field of a small magnet. The product of the current and the area of the loop is proportional to the moment of the magnet, i.e. to the product of pole strength and distance between the poles. The direction of the current is related to the magnet in such a manner that if the current flows anti-clockwise the nearer pole is a

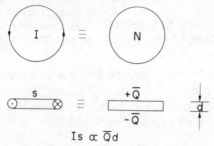

$$Is \propto \overline{Q}d$$

FIG. 5.4   *Equivalence of current loop and dipole*

north pole and if the current flows clockwise the nearer pole is a south pole." (See Fig. 5.4.) If it is asked how small is a small coil, the answer is that the coil and the magnet must be small in comparison to the distance at which the magnetic field is observed. The law does not allow us to investigate the field right on top of the coil or magnet. This restriction is similar to that in electrostatics where the law of force between charges does not work if we wish to discuss the field inside atoms.

Ampère believed that all magnetic effects were really due to current loops and he suggested that magnetic effects in iron were due to molecular currents. This was a remarkable suggestion to make at a time when there was no knowledge of atomic structure. The modern view is that magnetic effects inside materials are due to the orbital motion of electrons and, more importantly still, due to the spin of electrons. This explains why even at the subatomic level there are magnetic dipoles but no poles. The point is that magnetism is always due to rotating electric charges. Ampère's hypothesis was not far from the present view. To him, magnetism was due to the movement of electric charge, and he coined the word *electrodynamics* for the study of magnetic interactions. Nevertheless he realized that very often it was easier to solve problems by working in terms of magnetic poles rather than electric currents and for this purpose he invented a very ingenious weapon called the *magnetic shell*.

5.4. THE MAGNETIC SHELL AND THE CIRCUITAL LAW OF MAGNETISM

We wish to find the magnetic field due to a current loop. We know about poles and the inverse square law and we want to make use of this knowledge. How can we represent a current loop of any size by a set of magnetic dipoles?

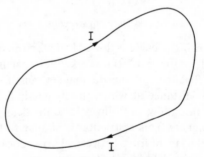

Fig. 5.5   *Large current loop*

The first law of electromagnetism tells us that the effect of a small current loop is the same as that of a small dipole. What about a large current loop? Ampère solved the problem in the following way. Figure 5.5 shows a current loop of arbitrary shape. Clearly it is possible to represent the current as being due to the small meshes of Fig. 5.6, because the current of the meshes will cancel out everywhere except at the boundary. The meshes can be made arbitrarily small and hence each one can be replaced by

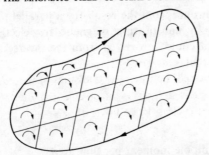

FIG. 5.6 *Large current loop divided into small loops*

a magnetic dipole. Then we shall have replaced the current loop by a double layer of magnetic poles. Ampère called this thing a magnetic shell.

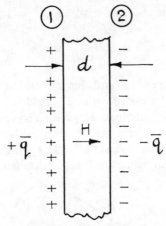

FIG. 5.7 *Magnetic shell*

Consider then an edge-on view of such a shell as shown in Fig. 5.7. Let the thickness of the shell be d and the pole strength $\bar{q}$ per unit area. The magnetic field strength inside the shell is given by

$$H = \frac{\bar{q}}{\mu_0} \tag{5.8}$$

This follows at once from the result for a parallel plate capacitor [eqn. (3.25)] by substituting magnetic for electric quantities. Alternatively it could be derived from the inverse square law or from Gauss's theorem.

The magnetic p.d. between the two sides of the magnetic shell is

$$V_2 - V_1 = \int_1^2 -H \, dl = -\frac{\bar{q}d}{\mu_0} \qquad (5.9)$$

and $\bar{q}d$ is the dipole moment per unit area.

$V_2 - V_1$ is independent of the path taken. Hence if instead of passing through the shell we take a path round the outside, $V_2 - V_1$ will have the same value as in eqn. (5.9). If we take a complete path which passes from one side of the shell through the shell to the other side and then back to the starting point without passing through the shell, we shall have

$$\oint H \, dl = 0I \qquad (5.10)$$

where the circle on the integral sign shows that the integration is carried out around a closed path. The field is conservative, there can be no change in potential energy in going round a closed path back to the starting point. Another way of looking at this is to say that the work done against the field by going round the outside is recovered by completing the journey through the shell.

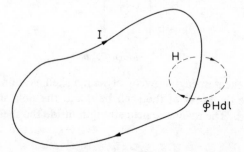

FIG. 5.8 *Circuital integration of* H *around current*

How can we apply this result to a current loop? Consider $\oint H \, dl$ in the two cases of the loop (Fig. 5.8) and the equivalent shell (Fig. 5.9). The only difference is that in the shell the energy

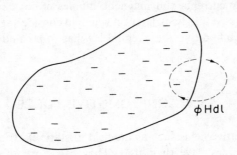

FIG. 5.9 *Circuital integration of* H *around magnetic shell*

is recovered by passing through the shell, whereas in the loop this energy is never recovered. Thus for the shell

$$\oint H \, dl = 0 \tag{5.10}$$

but for the loop

$$\oint H \, dl = \frac{\bar{q}d}{\mu_0} \tag{5.11}$$

and since the dipole moment is proportional to the current (by the first law of electromagnetism), we can choose the constant so that

$$\frac{\bar{q}d}{\mu_0} = I \tag{5.12}$$

and

$$\oint H \, dl = I \tag{5.13}$$

This is an equation of tremendous importance and value. Many writers pull it out of a hat and so dispense with the difficult argument about the magnetic shell. If you do not like the argument. just remember eqn. (5.13). But if you like a logical presentation remember that eqn. (5.13) is based on Ampère's equivalence of small current loops and magnetic dipoles. Another point to remember is that all our discussion has been concerned with steady currents and magnetic dipoles of constant strength. This will lessen the shock later on when, in considering alternating currents, we find that the circuital law has to be modified.

5.5. MAGNETOMOTIVE FORCE

Conservative fields such as the field of static electric charges or magnetic poles have their uses. They can store energy and so behave like springs. But if we need a system in which energy is to be converted steadily from one form to another, a conservative system is no use. A magnetic dipole layer is no better than an electrostatic capacitor in this respect.

However, a current loop is much better. The circuital law shows that a current loop is non-conservative. Every time a magnetic pole is carried around a complete loop enclosing the current, work is done equal to the current multiplied by the pole strength. Clearly this work must come from the source that is driving the current and we shall investigate this reaction more carefully in the next chapter.

Meanwhile we define the non-conservative $\int H \mathrm{d}l$ due to currents as *Magnetomotive Force*, by analogy with the electromotive force $\int E \mathrm{d}l$. There is, however, the slight practical difference that the electromotive force is often concentrated inside generators or batteries whereas the magnetomotive force is *distributed* in the space near the current.

If we want to be really careful we must now distinguish between H_1, the magnetic field strength due to magnets, and H_2, the magnetic field strength due to currents. Then

$$\oint H \, \mathrm{d}l = \oint (H_1 + H_2) \, \mathrm{d}l \qquad (5.14)$$

But $$\oint H_1 \, dl = 0 \tag{5.15}$$

because the field of magnets is conservative

Hence
$$\oint H \, dl = \oint H_2 \, dl$$
$$= I \tag{5.16}$$

Therefore the circuital law holds whether we use total magnetic field strength or only the component of magnetic field strength derived from electric current.

It is worth noting that the term magnetomotive force is as unfortunate as that of electromotive force. Magnetomotive force (m.m.f.) is not a force but an energy divided by the pole strength. It is also worth noting that we have to be careful about talking of magnetic potential difference. Potentials describe conservative fields and are therefore applicable to the fields of magnets and magnetic shells. We can use them even when the fields are those of current loops, because these loops can be replaced by shells. But this means that we must not apply the idea to any closed path linking a current, because it is the *closed* path which forces us to the distinction between H_1 and H_2, where H_1 is a potential gradient and H_2 an m.m.f. gradient.

5.6. APPLICATIONS OF THE CIRCUITAL LAW

The circuital law does not in general tell us what the magnetic field strength is at a point. In this respect it is similar to Gauss's theorem which only gives information about the total flux over a closed surface. Nevertheless we used Gauss's theorem to tell us the field strength at a point in certain important cases where we could invoke the symmetry of the structure. We shall use this trick again with the circuital law. It is a very common trick used by engineers. An actual problem is often far too difficult to solve, so instead we solve a simpler problem which is not very different.

The solution of the simple problem aided by common sense often goes far towards the solution of the actual problem. We make progress by shutting our eyes to the difficulties. It would of course be nice to be able to solve engineering problems in the way in which mathematical equations are solved, but this is generally impossible. The reader would be well advised to regard statements such as Gauss's theorem and the circuital law not as equations to be solved, but as statements giving the relationship between various physical quantities in a compact and easily remembered form.

5.6.1. *The Magnetic Field of a Long Straight Current*

Apply the circuital law around a circular path at a distance r from the axis of the wire (Fig. 5.10). By symmetry, H must be constant around this path.

$$\oint H\,\mathrm{d}l = H \times 2\pi r = I \qquad (5.17)$$

$$(5.18)$$

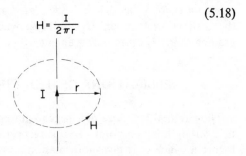

$$H = \frac{I}{2\pi r}$$

FIG. 5.10   *Magnetic field strength near long straight current*

Again, by symmetry, this is the only possible component of H and is thus the total magnetic field strength due to the current. Thus the magnetic field is disposed in circles around the wire. This was Oersted's experimental discovery.

Note the direction of the magnetic field, which can be obtained by considering a distant return conductor either to the right or to the left. It should be realized explicitly that the magnetic field has the direction of rotation of a right-handed screw when it is screwed in the direction of the current.

5.6.2. *The Magnetic Field Inside a Long Solid Cylindrical Conductor Carrying Uniformly Distributed Current*

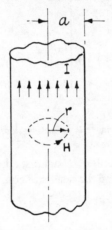

FIG. 5.11   *Magnetic field strength inside long cylindrical conductor*

Apply the circuital law around a circle at a distance r from the axis of the wire, where $r < a$, the radius of the conductor (Fig. 5.11). The current enclosed by the path will be

$$i = \frac{\pi r^2}{\pi a^2} I \qquad (5.19)$$

Hence

$$\oint H \, dl = H \, 2\pi r = \frac{r^2}{a^2} I \qquad (5.20)$$

and

$$H = \frac{r}{2\pi a^2} I \qquad (5.21)$$

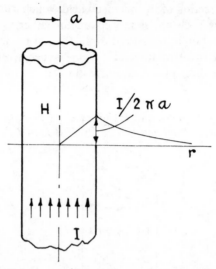

FIG. 5.12   *Complete magnetic field strength of cylindrical conductor*

Figure 5.12 shows a graph of H against r both inside and outside the conductor, combining the results of eqn. (5.18) and eqn. (5.21).

5.6.3. *The Magnetic Field Inside a Hollow Cylindrical Conductor Carrying Axial Current*

Apply the circuital law around a circle at a distance r from the axis of the tube, where $r < b$, the inner radius of the tube (Fig. 5.13).

$$\oint H\,dl = H\,2\pi r = 0 \qquad\qquad (5.22)$$

Hence $\qquad\qquad\qquad H = 0 \qquad\qquad\qquad\qquad (5.23)$

There is no magnetic field inside the tube.

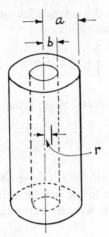

FIG. 5.13 *Magnetic field strength inside a hollow tube*

5.6.4. *The Magnetic Field Inside a Toroid*

Consider the ring shown in Fig. 5.14 both in plan and in section. Current is flowing around the surface of the ring as shown in the

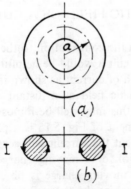

FIG. 5.14 *A toroid*

sectional view. Apply the circuital law around the centre line of the ring as shown in the plan view.

$$\oint H \, dl = H \, 2\pi a = I \tag{5.24}$$

where I is the *total* current flowing around the ring.

Hence
$$H = \frac{I}{2\pi a} \tag{5.25}$$

If the mean radius of the ring is large compared with the cross-section, the field in the toroid will be nearly uniform. It is usual to apply the current by winding many turns of wire around the ring. We then have

$$I = NI' \tag{5.26}$$

where N is the number of turns and I' is the current in each turn. Then

$$H = \frac{NI'}{2\pi a} \tag{5.27}$$

5.7. THE MAGNETIC FIELD OF A CURRENT ELEMENT

In order to obtain the contribution to the magnetic field made by each part of a circuit it would be helpful to find the magnetic field of a short length of current as shown in Fig. 5.15. We wish to obtain the magnetic field at a distant point P, but careful thought shows that this may well be impossible. The trouble is that the current element of Fig. 5.15 is a physical impossibility. When the circuit is cut the current will cease to flow. Nor is it any better if we think of a moving charge instead of a short current, although a moving charge is physically possible. The trouble is that we require the field of a steady current and this implies not a single moving charge but an endless succession of charges, so that straight away we are forced to consider the rest of the circuit. If we were thinking of a rapidly alternating current, a single oscillating charge might do, but this is not the case we are considering.

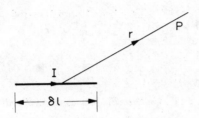

Fig. 5.15   *A current element*

This difficulty was overcome by Heaviside in a very ingenious manner. Suppose we think of a current element surrounded along its length by a thin insulating sheath—rather like a battery for an electric torch. The ends of the element are not insulated and the current is free to spread in all directions once it leaves the element,

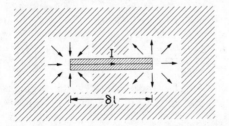

Fig. 5.16   *Heaviside's "rational" current element*

which is immersed in an ocean of conducting fluid (Fig. 5.16). Any circuit can be built up from such elements put end to end (Fig. 5.17), and when the last element has been put in position to close the circuit the conducting fluid can be drained away. Thus the fluid does not enter into any calculations on closed circuits, but its introduction enables us to calculate the contribution of each element of circuit.

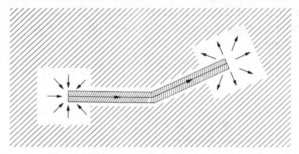

FIG. 5.17   *Two Heaviside elements end to end*

Consider then the magnetic field at P due to such a current element $I\delta l$ (Fig. 5.18). The flow of current is everywhere symmetrical about the axis of the current element. Hence the magnetic field must be disposed in circles around this axis. Consider the circuital law for a circle of radius R passing through P. What

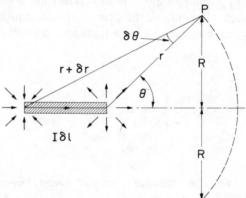

FIG. 5.18   *Magnetic field of a Heaviside current element*

is the current flowing through this circle? Consider first the outflow from the near end of the current element. This is the outflow over the spherical cap of radius r and angle 2θ at the centre. The area of such a cap is easily obtained (from a geometry book or by integration) and is found to be $2\pi r^2(1-\cos\theta)$. Thus the outflow of current is the current density multiplied by this area

$$i = \frac{I}{4\pi r^2} \times 2\pi r^2(1-\cos\theta)$$

$$= \frac{I}{2}(1-\cos\theta) \tag{5.28}$$

To obtain the net flow we must subtract the inflow through the same area towards the far end of the current element. This inflow can be obtained by inspection of eqn. (5.28). It is given by

$$i' = \frac{I}{2}(1-\cos[\theta-\delta\theta]) \tag{5.29}$$

Hence $\quad\quad i-i' = \frac{I}{2}(\cos[\theta-\delta\theta]-\cos\theta)$

$$= I \sin\theta \sin\frac{\delta\theta}{2}$$

$$= \frac{I}{2}\delta\theta \sin\theta \tag{5.30}$$

The circuital law gives

$$\oint H \, dl = H \, 2\pi R$$

$$= \frac{I}{2}\delta\theta \sin\theta \tag{5.31}$$

whence $\quad\quad H = \frac{I}{4\pi R}\delta\theta \sin\theta \tag{5.32}$

But $\quad\quad R = r \sin\theta \tag{5.33}$

and $\quad\quad r\,\delta\theta = \delta l \sin\theta \tag{5.34}$

whence $\quad\quad \mathbf{H} = \frac{I\,\delta l}{4\pi r^2}\sin\theta \tag{5.35}$

which is the expression for which we have been looking.

We note that once again we have an inverse square law, that the field is proportional to the current and the length and that its direction is perpendicular to the plane containing the element and the radius r. The "$\sin\theta$" multiplier is a little more difficult to remember, but it is clear that on the axis there can be no field (by symmetry) and this is a useful check when one has forgotten whether it should be $\sin\theta$ or $\cos\theta$.

5.8. USE OF THE FORMULA FOR THE MAGNETIC FIELD OF A CURRENT ELEMENT

5.8.1. *The Magnetic Field on the Axis of a Circular Current*

The circular current has radius a. We wish to find the magnetic field at a point P on the axis of the coil at a distance x from its centre (Fig. 5.19).

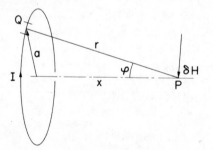

FIG. 5.19   *Magnetic field on the axis of a circular current*

The contribution to the magnetic field made by an element of current at Q is given by

$$\delta H = \frac{I\,\delta l \sin\theta}{4\pi r^2}$$

$$= \frac{I\,\delta l}{4\pi r^2} \qquad (5.36)$$

Since θ is 90° in this case.

The direction of δH is indicated in Fig. 5.19. Since P is on the axis, r is the same for all elements. The components of δH perpendicular to the axis will cancel, so that the total field is given by

$$H = \int \delta H \sin \phi \qquad (5.37)$$

Also $r \sin \phi = a$ (5.38)

whence $H = I \dfrac{2\pi a}{4\pi r^2} \dfrac{a}{r}$

$$= I \frac{a^2}{2r^3} \qquad (5.39)$$

and since $r^2 = a^2 + x^2$ (5.40)

$$H = I \frac{a^2}{2(a^2 + x^2)^{3/2}} \qquad (5.41)$$

5.8.2. *The Magnetic Field on the Axis of a Solenoid of Circular Cross-Section*

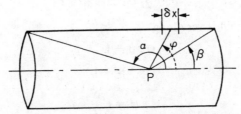

Fig. 5.20   *Magnetic field on the axis of a solenoid*

A solenoid can be built up from a set of coils each like that of Fig. 5.19. Let the current be i per unit length of solenoid. Then the contribution to the field H along the axis of the

solenoid by a typical section δx (Fig. 5.20) can be obtained from
eqn. (5.41)

$$\delta H = \frac{i \, \delta x \, a^2}{2(a^2 + x^2)^{3/2}} \qquad (5.42)$$

also
$$a = x \tan \phi \qquad (5.43)$$

whence
$$\delta x = -a \, \mathrm{cosec}^2 \, \phi \, \delta \phi \qquad (5.44)$$

therefore
$$\delta H = -\frac{i a^3 \, \mathrm{cosec}^2 \, \phi \, \delta \phi}{2 a^3 \, \mathrm{cosec}^3 \, \phi}$$

$$= -\frac{i}{2} \sin \phi \, \delta \phi \qquad (5.45)$$

whence
$$H = \frac{i}{2} [\cos \phi]_\alpha^\beta$$

$$= \frac{i}{2} (\cos \beta - \cos \alpha) \qquad (5.46)$$

In eqn. (5.46) i is the current per unit length. If a solenoid of
length l is wound with N turns carrying current I', we have

$$i = \frac{N}{l} I' \qquad (5.47)$$

and
$$H = \frac{I'N}{2l} (\cos \beta - \cos \alpha) \qquad (5.48)$$

If the solenoid is very long

$$\cos \beta = 1 \qquad (5.49)$$

and
$$\cos \alpha = -1 \qquad (5.50)$$

whence
$$H = I' \frac{N}{l} \qquad (5.51)$$

i.e. the magnetic field is equal to the ampere-turns per unit length,
a result which is in accordance with the result obtained for the
toroid eqn. (5.25). Clearly a very long solenoid has the same field
on its axis as a toroid of very large diameter.

5.9. THE FORCE ON A CURRENT ELEMENT IN A MAGNETIC FIELD

The magnetic field of a current element was calculated in Section 5.7 and is given by the expression

$$H = \frac{I\,\delta l}{4\pi r^2}\sin\theta \qquad (5.35)$$

Hence, if in Fig. 5.21 there is a magnetic pole of strength $+\bar{Q}$ at the point P, it will experience a force

$$F = H\bar{Q} = \frac{I\,\delta l \sin\theta\,\bar{Q}}{4\pi r^2} \qquad (5.52)$$

and this force will be out of the plane of the paper as indicated.

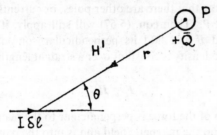

FIG. 5.21 *Reaction between a current element and a magnetic pole*

By Newton's third law there must be an equal and opposite reaction on the current element. There is no difficulty here about conducting oceans, because it is possible to measure the force on each separate element of a complete circuit.

Now let us express the reaction force of eqn. (5.52) in terms of the magnetic field of the pole. This field is given by

$$H' = \frac{\bar{Q}}{4\pi\mu_0 r^2} \qquad (5.53)$$

in the direction shown in Fig. 5.21.

Hence the force on the element can be written

$$F' = -F = I \, \delta l \, \mu_0 \, H' \sin \theta \tag{5.54}$$

Now $$B' = \mu_0 \, H' \tag{5.55}$$

Hence $$F' = I \, \delta l \, B' \sin \theta \tag{5.56}$$

But $B' \sin \theta$ is the component of B' perpendicular to the current element and so we can write it $B'_\perp$

Then $$F' = I \, \delta l B'_\perp \tag{5.57}$$

and dropping the dashes we have

$$F = I \, \delta l B_\perp \tag{5.58}$$

Thus the force is proportional to the current, the length of the element and the perpendicular component of magnetic flux density.

Suppose now that there are other poles, or currents, besides the pole at point P. Then eqn. (5.57) will still apply, if we take the *resultant* field B and find its perpendicular component. If the magnetic flux density is constant over a straight length l of current, we can write

$$F = IlB_\perp \tag{5.59}$$

The direction of the force is perpendicular to the plane containing the current and the magnetic field and is into the paper. This can be found by considering the force on the pole at P. Alternatively one can remember the sequence I, δl, $B_\perp$ and rotate δl until it is in line with $B_\perp$. The direction of the force is then the direction of motion of a right-handed screw. However, an easier way of remembering the direction will be given in the next section.

5.10. THE FORCE BETWEEN TWO LONG PARALLEL CURRENTS

The flux density at current I_2 due to current I_1 can be found as in Section 5.6.1 and is given by

$$B = \frac{\mu_0 I_1}{2\pi a} \tag{5.60}$$

perpendicular to I_2 (see Fig. 5.22). Hence the force per unit length of I_2 is

$$F = I_2 B$$

$$= \frac{\mu_0 I_1 I_2}{2\pi a} \tag{5.61}$$

Since B due to I_1 at the position of I_2 is into the paper, the force will be in a direction to bring the currents together (as explained in the last section), either by thinking of forces on poles or by moving from the direction of current flow into the direction of flux density and thinking of the motion of a right-handed screw.

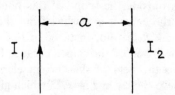

FIG. 5.22   *Two long parallel currents*

Another way of obtaining the same result which gives a far greater insight into the physical mechanism will now be described. Why is it that like currents as in Fig. 5.22 attract one another, whereas like charges and like poles repel? The answer was given by the great physicist Maxwell (about 1870). He pointed out that the energy associated with charges and poles is potential energy and that therefore these objects tend to move in a direction that will decrease the potential energy, just as a marble runs down an inclined plane. Now to reduce the potential energy is the same as to reduce the field which gives a measure of the potential energy. For instance if two like charges are brought together they strengthen one another's field, while opposite charges reduce one another's field. Thus like charges repel and unlike ones attract.

However, the energy associated with electric currents is not potential but kinetic in origin, because it is associated with moving charges. Now in mechanics it is well known that bodies move in a direction to increase their kinetic energy. For instance a marble accelerates down an inclined plane. The magnetic field of currents is a measure of kinetic energy and currents will try to move in a direction that will increase the field. Thus similarly-directed currents attract and unlike currents repel. The difference between kinetic and potential energy is easy to remember and it is a very powerful idea.

This method works well for the force between currents. Maxwell also provided a quick way of finding the direction of the force on a current in a magnetic field. He showed that the mathematical laws governing electric and magnetic fields are similar to the laws governing elastic bands. This is not difficult to prove but it is outside the scope of this book. The force is always such that the current is pushed from the side of strong field to that of weak field. For instance in Fig. 5.21 the field is stronger above the plane of the paper because here the field of the pole reinforces the field of the current element. The same result is shown more clearly in Fig. 5.23 which gives the force on a current in a magnetic field. To find the direction of the force we add the field of the current to the field of the other sources. Hence the force is into the paper in Fig. 5.21 and vertically upwards in Fig. 5.23. We can summarize the rules for the direction of the force as follows:

Because the magnetic field of currents is a measure of kinetic energy, the forces on currents are always in a direction to increase the magnetic field. Thus like currents attract and unlike ones repel. If it is required to find the direction of the force on a current in a magnetic field, one can either replace the field by its current sources and find the interaction between the currents, or

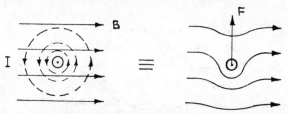

FIG. 5.23 *Force on a current in a magnetic field*

more simply one can say that the force is perpendicular to current and field in a direction from the region of stronger field to that of weaker field.

5.11. THE FORCE ON A COIL IN A MAGNETIC FIELD

Consider a coil (Fig. 5.24) which carries a current I in an arbitrary magnetic field B. We wish to know the force on this coil in a particular direction, which can be taken as the x axis.

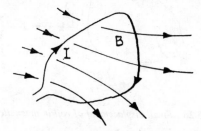

FIG. 5.24 *Coil in non-uniform magnetic field*

If the field B is parallel to the x axis, as in Fig. 5.25, there can be no force in this direction because we know that the force on each element of the coil is perpendicular to the magnetic field at the element. There is a magnetic flux linking the coil in the x

direction, but this does not produce a force in that direction. However, in Fig. 5.24 there are components of B in directions perpendicular to the x axis and these may produce a resultant

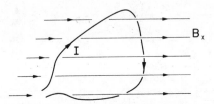

FIG. 5.25   *Coil in uniform magnetic field*

force along this axis. We shall approach the problem by considering a small displacement of the coil in the x direction (Fig. 5.26). Let us apply Gauss's theorem to the box formed by the coil in its original position and in its new position. The edge of the box is formed by the ribbon of width δx. If the flux entering the box on the left hand side is Φ, then in mathematical language the

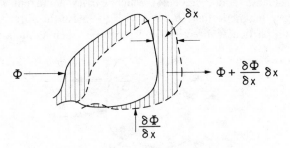

FIG. 5.26   *Small displacement of coil in magnetic field*

flux coming out of the right hand side is Φ plus the change of Φ, i.e. $\Phi + (\partial\Phi/\partial x)\delta x$. That means that a flux of $(\partial\Phi/\partial x)\delta x$ must enter the box around the ribbon. Of course $\partial\Phi/\partial x$ can be positive or negative and in the case of Fig. 5.24 it is actually negative. Now let the perimeter of the coil have length l and let the flux density perpendicular to the ribbon be $B_\perp$ ($B_\perp$ need not be constant along l).

Then

$$\delta x \oint B_\perp \, dl = \frac{\partial \Phi}{\partial x} \delta x \qquad (5.62)$$

Therefore

$$\oint B_\perp \, dl = \frac{\partial \Phi}{\partial x} \qquad (5.63)$$

Now the force in the direction of x on any element of the perimeter is

$$\delta F_x = I \, \delta l B_\perp \qquad (5.64)$$

Hence the total force is given by

$$F_x = I \oint B_\perp \, dl$$

$$= I \frac{\partial \Phi}{\partial x} \qquad (5.65)$$

If there are N closely wound turns on the coil

$$F_x = NI \frac{\partial \Phi}{\partial x} \qquad (5.66)$$

This is a very important result. The reader should notice explicitly that the force depends on the rate of change of flux in the direction of the force. It does not depend on the flux linking the coil.

Similarly we can derive

$$F_y = I \frac{\partial \Phi}{\partial y} \qquad (5.67)$$

$$F_z = I \frac{\partial \Phi}{\partial z} \qquad (5.68)$$

and

$$T_\theta = I \frac{\partial \Phi}{\partial \theta} \qquad (5.69)$$

where T_θ is the torque trying to move the coil through the angle θ.

The sign of the force can be found either by the right-handed screw rule or by replacing the flux by its source in the form of another coil. For instance the flux in Fig. 5.24 could be due to a coil with a current flow in the same sense as the coil shown and placed to the left of this coil. Since like coils attract, the force on the coil shown will therefore be towards the left. It will have been noticed that in this section we have not considered the flux due to I and the reader may question whether this is correct. Actually, it does not matter. The flux due to I does not change as the coil moves. Hence it does not contribute to the force. This is reasonable, since otherwise a coil might move purely under the action of its own flux and such behaviour would be that of a perpetual-motion machine.

5.11.1. *The Force Between Coaxial Circular Currents*

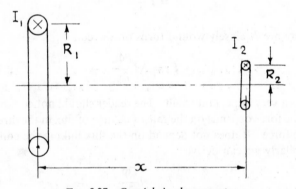

FIG. 5.27   *Coaxial circular currents*

Let the radii be R_1 and R_2 where $R_1 \gg R_2$ and let the currents be I_1 and I_2 (Fig. 5.27). Then

$$F_x = I_1 \frac{\partial \Phi_2}{\partial x} = I_2 \frac{\partial \Phi_1}{\partial x} \qquad (5.70)$$

where Φ_1 is the flux due to current I_1 linking the coil carrying current I_2 and Φ_2 is the flux due to I_2 linking I_1. Since $R_1 \gg R_2$ we can assume that Φ_1 is uniformly distributed over the area of coil 2. This means that it is very much easier to calculate Φ_1 than Φ_2.

$$\Phi_1 = \pi R_2^2 B_1 = \pi R_2^2 \mu_0 H_1 \tag{5.71}$$

and
$$H_1 = \frac{I_1 R_1^2}{2(R_1^2 + x^2)^{3/2}} \tag{5.72}$$

a result found in eqn. (5.41). (There is of course no need to memorize this result. We are merely trying to avoid repetition.)

Hence
$$\Phi_1 = \frac{\mu_0 I_1 \pi R_1^2 R_2^2}{2(R_1^2 + x^2)^{3/2}} \tag{5.73}$$

and
$$F_x = -\tfrac{3}{2}\pi\mu_0 \frac{I_1 I_2 R_1^2 R_2^2 x}{(R_1^2 + x^2)^{5/2}} \tag{5.74}$$

This is an attractive force if the currents have the same sense.

5.11.2. *The Dynamometer Wattmeter*

One form of this instrument is illustrated in Fig. 5.28. A small moving coil is placed half-way between two large fixed coils. The

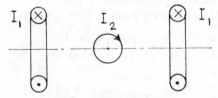

Fig. 5.28 *Dynamometer coils*

moving coil carries a current proportional to the potential difference of the circuit whose power is to be measured, while the fixed coils carry the current in the circuit. The torque on the moving coil is given by

$$T = I_2 \frac{\partial \Phi_1}{\partial \theta} \qquad (5.75)$$

Now $I_2 \propto V$ and $\Phi_1 \propto I_1$

Hence $\qquad\qquad T = KVI_1 \qquad (5.76)$

and thus the instrument measures watts. It should be noted that the maximum torque is obtained when $\partial \Phi_1/\partial \theta$ is a maximum, i.e. when the plane of the moving coil is perpendicular to the planes of the fixed coils. In this position there is no flux linkage between the fixed coils and the moving coil. Once again it is not the flux linkage that matters, but the rate of change of flux linkage. The detailed analysis of the wattmeter torque follows closely on the analysis of Section 5.11.1.

5.12. MAGNETIC UNITS

In Chapter 2 we defined the unit of electric charge as the coulomb or ampere-second. Thus in electrostatics we had to accept a unit which had been previously defined. We now come to this definition which has been made by international agreement as follows: "The ampere is that constant current which, if maintained in two straight parallel conductors of infinite length, of negligible circular cross-section, and placed 1 metre apart in vacuum, would produce between these conductors a force equal to 2 × 10⁻⁷ newton per metre of length."

Equation (5.61) gives the formula for this force per unit length as

$$F/m = \frac{\mu_0 I_1 I_2}{2\pi a} \qquad (5.61)$$

Thus we have

$$2 \times 10^{-7} = \frac{\mu_0}{2\pi} \frac{1 \times 1}{1} \qquad (5.77)$$

in the S.I. system. Hence in this system

$$\mu_0 = 4\pi \times 10^{-7} \qquad (5.78)$$

This definition defines all other magnetic quantities. The unit of pole strength is defined by eqn. (5.1), the unit of magnetic field strength by eqn. (5.2), magnetic potential by eqn. (5.3), flux density by eqn. (5.4), flux by eqn. (5.5) and magnetomotive force in Section 5.5. Thus since all magnetic and electrostatic units are defined once the ampere is defined, we have shown that one electrical quantity in conjunction with the mechanical units of length, mass and time is sufficient to give a consistent set of electromagnetic units.

Since $\oint H \mathrm{d}l = I$, the magnetic field strength has dimensions of ampere/metre. The unit of flux is called the weber, and by Gauss's theorem pole strength can also be measured in webers. The dimensions of the weber can be obtained as follows:

$\int H \mathrm{d}l$ is (work done/pole strength). Thus dimensionally

$$\text{amperes} = \text{joules/webers}$$

and $$\text{webers} = \frac{\text{joules}}{\text{amperes}} = \frac{\text{watt-seconds}}{\text{amperes}} = \text{volt-seconds}$$

An alternative way in terms of the four primary quantities is to write

$$\text{webers} = \frac{\text{joules}}{\text{amperes}} = \frac{\text{kg m}^2}{\text{s}^2 \text{ A}} \quad *$$

The dimensions of μ_0 are given by

$$\mu_0 = \frac{B}{H} = \frac{\text{weber/m}^2}{\text{amp/m}} = \frac{\text{weber}}{\text{amp-metre}} = \frac{\text{Wb}}{\text{Am}}$$

or in terms of the four primary quantities

$$\mu_0 = \frac{\text{kg m}^3}{\text{s}^2 \text{ A}^2}$$

The units of magnetomotive force and magnetic potential difference have the dimension ampere. The unit of magnetic flux density is the tesla (T). Thus 1 weber/m^2 is 1 tesla.

---

* In the next chapter we shall see that this is henry/metre, where the henry is the unit of inductance.

SUMMARY

In Chapter 5 we have begun to examine the relationship between electricity and magnetism. We have stated the first law of electromagnetism which shows the equivalence of current loops and dipoles. This has led us to the idea of magnetic shells and by the use of this idea we have been able to derive the circuital law of magnetism.

By means of the circuital law we have been able to calculate the magnetic field of a current element and also the force on a current element in a magnetic field.

Finally we have shown that a consistent system of electromagnetic units can be defined by using one electrical entity in addition to the three mechanical entities of mass, length and time. The entity chosen by international agreement is electric current and its unit, the ampere, is defined in terms of force between two long currents.

NEW TERMS USED IN THIS CHAPTER

Term	Symbol	Unit (Abbreviation)	Definition
Magnetic Pole	$\bar{Q}$	weber (Wb)	$F = \dfrac{\bar{Q}_1 \bar{Q}_2}{4\pi\mu_0 r^2}$
Magnetic Field Strength	H	ampere/metre (A/m)	$H = \dfrac{\bar{Q}}{4\pi\mu_0 r^2}$
Magnetic Potential Magnetic Potential Difference	V	ampere (A)	$V_2 - V_1 = \displaystyle\int_1^2 -H\,\mathrm{d}l$
Magnetic Flux Density	B	tesla (T)	$B = \mu_0 H^*$
Magnetic Flux	Φ	weber (Wb)	$\Phi = \displaystyle\iint_s B_n\,\mathrm{d}S$
Pole Surface Density	$\bar{q}$	tesla (T)	$\bar{Q} = \displaystyle\iint_S \bar{q}\,\mathrm{d}S$
Magnetomotive Force	m.m.f.	ampere	$\displaystyle\oint H\,\mathrm{d}l = I$
Magnetic Constant	μ_0	henry/metre (H/m)	$F/m = \dfrac{\mu_0 I_1 I_2}{2\pi a}$

Note: The weber can also be written volt-second.

* In free space.

Exercises

5.1. Obtain the expression for the magnetic field of a long straight current [eqn. (5.18)] by using the formula for the field of a current element and integrating.

5.2. A coil is wound with 10 closely concentrated turns in the form of a square of side 16 cm. Find the magnetic field strength at a point on the axis of symmetry of the coil distant 15 cm from its plane caused by a current of 5 A in the coil.
(*Ans.* 37·4 A/m)

5.3. A closely wound coil of 80 turns is made in the form of a regular hexagon with sides 12 cm long and carries a current of 2 A. Find the magnetic field strength at a point on the axis of symmetry 5 cm from the plane of the coil.
(*Ans.* 550 A/m)

5.4. Define *magnetic field strength* and *magnetic flux*. A solenoid is uniformly wound with 180 turns of fine wire; the diameter of the solenoid is 60 mm and its length is 100 mm. A concentrated circular coil of 20 mm diameter, wound with 100 turns of fine wire, is placed coaxially and centrally within the solenoid. Estimate the flux linkage with the coil when there is a current of 500 mA in the solenoid.
(*Ans.* 30·5 × 10^{-6} Wb)

5.5. Show that the magnetic flux density on the axis of a long solenoid is approximately twice as strong at the centre as at one of the ends. Give values for the ratio of length/diameter of the solenoid for this result to be true within (a) 10%, (b) 1%.
(*Ans.* 1·7, 6·4)

5.6. Find an expression for the magnetic field strength at a point P on the axis of a coil of N turns of radius R cm, closely wound and carrying a current I amperes, at a distance x from the plane of the coil.

If a second closely wound coil of n turns of radius r cm (r being much less than R) is mounted at P, being pivoted about a diameter parallel to the plane of the larger coil and carrying a current of i amperes, show that the torque on the small coil about its axis of suspension is approximately

$$\frac{2\pi^2 R^2 r^2 Nn I i}{(R^2 + x^2)^{3/2}} \sin \theta \times 10^{-9} \text{ Nm}$$

where θ is the angle between the planes of the coils.

5.7. State and comment on the rules for finding (a) the magnitude and direction of the magnetic field strength at a point due to a current in an element of a neighbouring conductor, and (b) the mechanical force on an element of a current-bearing conductor lying in a magnetic field. Show that these rules are mutually compatible.

5.8. Two equal circular coils are coaxial, the distance between the planes of the coils being equal to the radius of either. The coils are connected in series so that the magnetic fields caused by current in them have the same polarity. Show that the magnitude of the magnetic flux density at a point on the axis of the coils and mid-way between them is about 5·7% greater than that at the centre of either coil.

5.9. Define *magnetic pole strength* and find its dimensions. Two closely wound circular coils can be mounted co-axially. Coil 1 has a mean diameter of 250 mm, 100 turns and carries 5A. Coil 2 has a mean diameter of 15 mm, 500 turns and carries 4 mA. At what distance must the coils be placed so that the force between them is a maximum and what is the value of this force?

(*Ans.* 62·5 mm, 6·1 × 10⁻⁶N)

5.10. Distinguish between magnetomotive force and magnetic potential difference. Under what circumstances is it possible to represent the magnetic field of an electric current in terms of a magnetic potential?

5.11. Derive from first principles an expression for the force on a current-carrying coil in a magnetic field. Does your expression hold for coils of arbitrary shape or only for plane coils?

CHAPTER 6

Electromagnetic Induction

6.1. THE MOTION OF ELECTRIC CHARGES THROUGH MAGNETIC FIELDS

In the last chapter we began to investigate the relationship between electricity and magnetism and we saw that magnets and currents produce forces on one another. As a result of our discussion we now know a great deal about the relationship between magnets and currents, but this knowledge is still incomplete, because so far we have not discussed the *motion* of magnets or currents, nor have we allowed the currents to vary in strength. Moreover the last chapter has been unsatisfactory inasmuch as we have said much about forces on currents, but nothing about forces on electric charges, although it is the object of this book to deal with the statics and dynamics of electric charge. In fact the last chapter was really a digression which was

FIG. 6.1 *Equivalence of moving charges and current*

needed to introduce the reader to the idea of the magnetic field and to the various terms which are used in the description and analysis of magnetic fields. In this chapter we return to the main stream of the argument. First of all we must reinterpret the force exerted on a current in a steady magnetic field in terms of the force on a moving charge. Figure 6.1 shows a succession of

120

electric charges Q spaced at a distance δl moving with steady velocity v. The current is the rate of passage of charge, so that we can write

$$I = \frac{Q}{\delta t} \qquad (6.1)$$

and since
$$\delta l = v\,\delta t \qquad (6.2)$$

$$I = \frac{Qv}{\delta l} \qquad (6.3)$$

We can now regard the average current I over a length δl as being due to the motion of a single charge Q and we can use eqn. (6.3) to state the equivalence

$$I\,\delta l = Qv \qquad (6.4)$$

which means that a current element is equivalent to a charge multiplied by its velocity.

Now the force on a current element was given in the previous chapter as

$$F = I\,\delta l B_\perp \qquad (5.57)$$

so we can now write

$$F = QvB_\perp \qquad (6.5)$$

and we have made the important deduction that an electric charge moving with a velocity v through a magnetic field B experiences a force which is proportional to the charge, the velocity and the component of flux density perpendicular to the velocity. The force acts perpendicularly to the plane containing the velocity vector and the magnetic flux density. The sign of the force can be obtained from the rule deduced for electric current (see Fig. 6.2).

This is undoubtedly one of the most important results in the whole of electrical science. It is far too important a result to be based on a process of reasoning alone. It needs to be verified experimentally. The reader should take the first possible opportunity to satisfy himself in the laboratory that moving electric charges can be deflected by magnetic fields. One way of doing this is to study the behaviour of a cathode ray oscillograph fitted with magnetic deflection coils or a magnetic focusing coil. Notice especially that the charges have to move *across* the magnetic field. There is no force on a charge moving in the direction of the magnetic field.

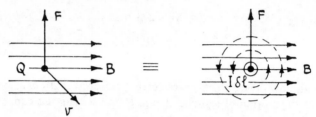

FIG. 6.2   *Force on moving charge in magnetic field*

It will have been noticed that in the last chapter we have talked somewhat loosely about forces on currents and forces on conductors. We are now able to see the matter more clearly. The force on a conductor carrying a current in a magnetic field arises as follows. The moving charges experience a force in accordance

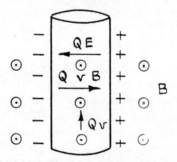

FIG. 6.3.   *Conductor in a magnetic field*

with eqn. (6.5). This results in a displacement of charge and a surface charge appears as shown in Fig. 6.3. Since the charges cannot escape from the surface, they exert a pull on the stationary lattice of the conductor and in this way the force is transferred to the conductor. The transverse movement of charge to the surface ceases when the electrostatic force from the surface charges cancels the motional force inside the conductor. The surface charges produce a potential difference which can be measured with a voltmeter and is called the *Hall-effect voltage* after its discoverer.

6.1.1. The Motion of an Electron Which Has Been Given a Velocity v Across a Magnetic Field B

The force on the electron will be

$$F = evB \tag{6.6}$$

where e is the electronic charge. This force is perpendicular to both v and B. Since the force is always perpendicular to v, the magnitude of v will not change although its direction will change. Moreover since the force is constant the electron will describe a circle of radius r given by the equation

$$evB = \frac{mv^2}{r} \tag{6.7}$$

where m is the mass of the electron. Hence the angular velocity

$$\omega = \frac{v}{r} = \frac{eB}{m} \tag{6.8}$$

Thus the electron will describe circles with angular velocity $(e/m)B$ which is independent of v. This fact is used in a cathode-ray tube which is provided with a magnetic focusing coil (Fig. 6.4). Electrons emerge from the hole through the anode with varying

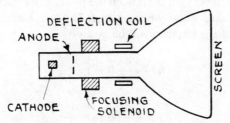

FIG. 6.4 *Cathode-ray tube with magnetic focusing and deflection coils*

transverse velocities, but their angular velocity in the magnetic field will be the same. Hence this field can be arranged to make every electron travel around a complete circle in its spiral path towards the screen. This will mean that all the electrons will arrive at the same spot on the screen. This device can be used to focus a beam of electrons.

6.1.2. *The Motion of an Electron Under the Combined Action of Electron and Magnetic Fields*

In a number of electronic devices, such as the magnetron, charges are forced to move under the action of both electric and magnetic fields. Let us consider the simple case shown in Fig. 6.5 in which an electron moves between two charged parallel

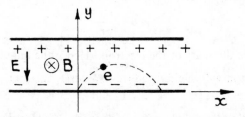

Fig. 6.5 *An electron moving through crossed electric and magnetic fields*

plates. There is an electrostatic field E as shown and also a uniform magnetic field B between the plates in a direction into the paper. Let axes x and y be chosen as shown. Then the equation of motion of the electron will be given by

$$Ee - e\frac{\mathrm{d}x}{\mathrm{d}t}B = m\frac{\mathrm{d}^2y}{\mathrm{d}t^2} \tag{6.9}$$

and
$$e\frac{\mathrm{d}y}{\mathrm{d}t}B = m\frac{\mathrm{d}^2x}{\mathrm{d}t^2} \tag{6.10}$$

it being remembered that electrons have negative charge. Integrating eqn. (6.10)

$$eyB = \frac{m\,dx}{dt} + C \tag{6.11}$$

And if $y = 0$ when $dx/dt = 0$ then $C = 0$. Substituting in eqn. (6.9) we obtain

$$m\frac{d^2y}{dt^2} + \frac{e^2B^2}{m}y = Ee \tag{6.12}$$

which has the solution

$$y = \frac{Em}{eB^2}\left(1 - \cos\frac{eBt}{m}\right) \tag{6.13}$$

if $y = 0$ and $dy/dt = 0$ when $t = 0$. Also from eqn. (6.11)

$$x = \frac{E}{B}\left(t - \frac{m}{eB}\sin\frac{eBt}{m}\right) \tag{6.14}$$

The electron will therefore move in a curve like that indicated in Fig. 6.5.

6.2. ELECTROMOTIVE FORCE INDUCED IN A CONDUCTOR MOVING THROUGH A MAGNETIC FIELD

The force on a charge moving across a magnetic field is given by

$$F = QvB \tag{6.15}$$

Hence we can define an electric field strength

$$E = vB \tag{6.16}$$

which is force/charge.

In other words, motion across a magnetic field produces an electric field. The relationship of the three vectors E, v and B is shown in Fig. 6.6.

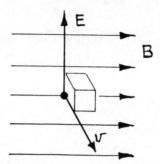

FIG. 6.6   *Motional electric field strength*

It will be remembered that in magnetism we distinguished two types of magnetic field strength: H_1 the magnetic field strength due to magnets and H_2 the magnetic field strength due to currents. H_1 describes a conservative field and is associated with magnetic potentials and H_2 describes a non-conservative field and is associated with magnetomotive force. We now have a parallel situation in electricity. The electrostatic field strength E_1 is due to electric charges and is associated with a conservative field and potential, while the motional electric field strength of eqn. (6.16) is associated with a non-conservative field and electromotive force. Just as the total magnetic field strength is the sum of H_1 and H_2, so is the total electric field strength experienced by a charge the sum of the two components E_1 and E_2.

$$E_{\text{total}} = E_{\text{electrostatic}} + vB \qquad (6.17)$$

In Section 4.4 we pointed out that an electromotive force is necessary to circulate electric current and that most generators rely for their electromotive force on the motion of conductors through magnetic fields. Let us consider this process in detail.

Figure 6.7 shows a conductor of length l moving with velocity v across a uniform magnetic field B. The charges in the conductor will experience an electric field strength vB and the free electrons

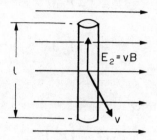

FIG. 6.7   *Motional electric field strength in a conductor*

will move as far as they can, i.e. a surface charge will appear on the conductor (Fig. 6.8). When the electrons cease to move, the amount of the surface charge will be such as to provide an electrostatic field strength which is equal and opposite to vB. The motion of the charges will constitute a very small transient current and will take a negligible time. Once the transient current has died away, no further current flows in the conductor.

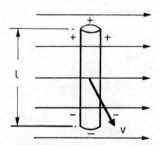

FIG. 6.8   *Electrostatic field cancels motional field*

The surface charges provide a potential difference which can be measured by a stationary voltmeter connected to the moving conductor through sliding contacts. By producing such a potential difference the moving conductor behaves like a chemical battery. The potential difference will be

$$V = \int -E_1 \, \mathrm{d}l$$

$$= \int +E_2 \, \mathrm{d}l$$

$$= vBl \qquad (6.18)$$

The electromotive force will be

$$\text{e.m.f.} = \int E_2 \, \mathrm{d}l$$

$$= vBl \qquad (6.19)$$

Thus the e.m.f. and the p.d. are equal.

So far the moving conductor does not supply current, because there is no circuit, but now consider the arrangement of Fig. 6.9 where there is a closed circuit. A thick conductor slides with

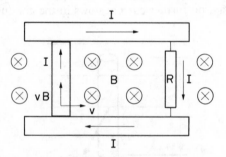

FIG. 6.9 *A simple generator*

velocity v between two thick conducting bars through a magnetic field B. A p.d. of $V = vBl$ will be set up between the upper and lower bar. This p.d. will be due to the surface charges which have been displaced by the motional electric field strength vB in the conductor. These are not shown in Fig. 6.9, but they should not be forgotten.

The p.d. will cause a current I to flow through the resistor R in accordance with Ohm's law

$$I = V/R \qquad (6.20)$$

and the current will return through the rest of the circuit as indicated. Electric energy will be converted to heat in the resistor at the rate of VI watts. This energy comes from the moving conductor which supplies it at the rate

$$\text{e.m.f.} \times I = VI \qquad (6.21)$$

We have assumed that the moving conductor and the conducting bars have no electrical resistance, i.e. that the current flows through them without requiring any force to drive it. Thus we have assumed that the electric field strength in the moving conductor is zero with the electrostatic field just cancelling the

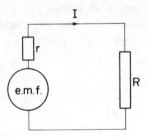

FIG. 6.10 *The equivalent circuit of the generator*

motional field. This is of course only an approximation because there will always be some loss of energy in the generator. Thus the motional vB must always be slightly greater than the electrostatic E_1, i.e. the e.m.f. must always be slightly greater than the p.d. One way of describing this is by the equivalent circuit of Fig. 6.10, where r is the internal resistance of the generator.

We have seen that the electrical energy comes from the moving conductor. How then is this energy supplied to the conductor?

Since the generator only works when it is moving, the answer must surely be that the energy is supplied by the device that moves the conductor. Consider then the force on the moving conductor.

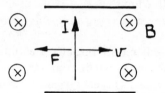

FIG. 6.11   *Force on a conductor moving through a magnetic field*

This force is given by $F = IlB$ and its direction is opposite to v (Fig. 6.11). Thus work has to be done on the conductor to keep it in motion. The rate of working is

$$Fv = IlBv$$

$$= I \times \text{e.m.f.} \tag{6.22}$$

Thus mechanical energy is converted to electrical energy. It is by means of this principle that the overwhelming proportion of electrical energy is generated all over the world.

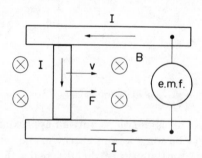

FIG. 6.12   *A simple motor*

It is interesting to enquire whether this process of energy conversion is reversible. Suppose we disconnect the resistor in Fig. 6.9 and connect instead a generator as in Fig. 6.12. If the e.m.f.

of the generator is larger than vBl the direction of current will be reversed, and the force F will also be reversed and will act in the direction of motion. Thus the moving conductor could drive a mechanical device, for instance a machine tool. The reversal of current flow has turned our generator into a motor. Clearly the process of energy conversion from mechanical to electrical energy is reversible. It should be noted that the efficiency of the conversion process is likely to be high. Losses will be due to internal resistance and friction. In fact efficiencies of 98 % are possible for large generators and motors. The efficiency of smaller machines will be a few per cent less.

Real generators and motors look very different from the primitive devices of Figs. 6.9 and 6.12. Most of them rotate and so provide continuous motion. They also employ many conductors instead of just one. The subject of electrical machines is outside the scope of this book, but it is an important and fascinating one and it forms the basis of a vast industry.*

Although we have based our discussion on the motion of a solid conductor, the same principles apply to conducting liquids or gases. The case of conducting liquids is important in the pumping of liquid metals and the case of conducting gases is receiving a lot of attention, because it is hoped to generate electricity by passing hot conducting gases through magnetic fields. This process is often referred to as the *direct generation* of electricity.

6.3. ELECTROMOTIVE FORCE INDUCED IN A STATIONARY CIRCUIT BY A CHANGING MAGNETIC FIELD

The discussion of generators and motors in this chapter has been based on the force experienced by a charge moving through a constant magnetic field. The existence of this force was deduced by applying Newton's third law of action and reaction to the case

* See, for instance, J. Hindmarsh: *Electrical Machines and their Applications*, a companion volume in this series.

of a moving charge and a magnet, an effect which could have been deduced by Ampère from his first law of electromagnetism. However, motional electromotive forces were in fact discovered by Faraday some ten years after Ampère's great work. Under Faraday's influence the subject of electromagnetism made tremendous strides and it is well worth while to look at the way

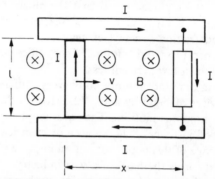

FIG. 6.13   *Generator supplying current to a resistance*

in which he approached it. Faraday did not think of electromotive force as being due to the force on electric charges. He dealt only with closed circuits and concentrated his attention on the magnetic flux linking these circuits. Consider Fig. 6.13 which shows again the simple generator discussed in the last section. The flux linking the circuit is given by

$$\Phi = \iint_S B \, dS = Blx \qquad (6.23)$$

Now the e.m.f. $= vBl$ by eqn. (6.19) and

$$v = -\frac{dx}{dt} \qquad (6.24)$$

Thus we can write

$$\text{e.m.f.} = -\frac{dx}{dt} Bl = -\frac{d\Phi}{dt} \qquad (6.25)$$

The electromotive force acting around the circuit is equal to the rate of change of the flux linked with the circuit. This is what

Faraday found, and his discovery led him to tremendous further discoveries.

Before we follow Faraday a word should be said about the negative sign in eqn. (6.25). The reader will rightly feel that this sign is somewhat arbitrary and depends on the positive directions we care to choose for v and B. This is true, but it is nevertheless useful to retain the sign for the following reason. If we examine the magnetic field component due to the current I being generated we find that it is always in such a direction as to *oppose* the change of the magnetic field which induces the e.m.f. Thus the motion in Fig. 6.13 reduces the flux in the circuit, but the current strengthens it. This is very reasonable. If the generated current *aided* the change of magnetic field the process would become unstable. The fact that the process is stable and that the current opposes the change in magnetic field is called *Lenz's Law* and we remind ourselves of the law by putting the negative sign in eqn. (6.25).*

Now we return to Faraday's researches. Equation (6.25) was derived from the fact that a charge moving through a magnetic field experiences a force. Suppose now the charge is stationary and the field moves, what happens then? But what does it mean to say that the field moves? The field is a mathematical symbol not a material object. Suppose we are more careful and say that the sources of the field, which may be magnets or currents, move. Does eqn. (6.25) still hold? Faraday found that it did. Now suppose further that there is no motion of magnets and currents, but that the flux changes because the currents which produce the flux change in magnitude. Does eqn. (6.25) still hold? Again Faraday found that it did.

If eqn. (6.25) had been merely a restatement of the result that a charge moving through a magnetic field experiences a force, it would have been of small interest. What is so revolutionary about it is that eqn. (6.25) applies not only to the special case of charges moving through constant magnetic fields but to the general case of circuits of arbitrary shape moving or stationary in constant or changing magnetic fields, and that the change in these magnetic

* If we reverse the direction of the arrow for I and think of a current that is being fed into the moving conductor, a positive sign is needed in eqn. (6.25).

fields can be due to motion or to a change in strength of the sources of the field. It is this generality that has made eqn. (6.25) one of the most useful relationships in electromagnetism. It is the *second law of electromagnetism* and as a tribute to its inventor it is generally known as Faraday's law of electromagnetic induction. It is interesting to note that it was a consideration of this law that led Einstein to his theory of relativity.

A record of Faraday's researches exists in his diary and this makes fascinating reading, but once again it must be emphasized that electrical engineering cannot be learned from books only. The reader should test experimentally whether Faraday's law is correct.

Equation (6.25) applies to the e.m.f. in a single loop. If instead of a single loop we have a coil of N turns, then the flux is linked N times and the e.m.f. is N times as large. We can thus write

$$\text{e.m.f.} = -N\frac{d\Phi}{dt} \tag{6.26}$$

This equation is sometimes stated loosely by saying that the electromotive force is equal to the rate of change of flux turns. A more adequate statement would be that the electromotive force is equal to the turns multiplied by the rate of change of flux. There is no e.m.f. if the turns around the flux are changed and the flux is kept constant.

The causes of e.m.f. summarized in Faraday's law are two:

(1) An e.m.f. is produced if a part of a circuit moves across a magnetic field.

(2) An e.m.f. is produced in a stationary circuit if the flux linked with the circuit changes.

One further point needs to be made. How can we interpret Faraday's law in terms of forces on electric charges? So far we have observed two types of such forces: electrostatic forces and motional forces. Since Faraday's law deals with closed circuits and since the integral of electrostatic field strength around a closed circuit is zero, the law tells us nothing about the presence or absence of electrostatic forces. The law does tell us something about the motional forces and it also tells us that there are additional forces due to changing magnetic fields. However, the law deals only with e.m.f., i.e. with the integral of electric field strength around the closed circuit. We are not told whereabouts in the circuit these forces act nor can we differentiate between the two types of e.m.f., that due to motion and that due to change of magnetic field. The law deals only with the total e.m.f.

It would be useful to be able to write

$$E_{total} = E_1 + E_2 + E_3 \qquad (6.27)$$

where E_1 is electrostatic field strength, E_2 is vB and E_3 is due to a change of magnetic field with time (a $\partial B/\partial t$ effect). Equation (6.27) is indeed used a great deal, but the expression for E_3 is not simple and is beyond the scope of this book.*

If we rewrite Faraday's law in terms of eqn. (6.27) we have

$$\text{e.m.f.} = \oint (E_2 + E_3)\,dl$$

$$= \oint (vB + E_3)\,dl = -\frac{d\Phi}{dt} \qquad (6.28)$$

6.3.1. Application of Faraday's Law: A Simple Transformer

We have seen in Section 6.2 how a motional e.m.f. is used by electrical engineers to convert mechanical into electrical energy and vice versa. Now we wish to consider briefly the application of Faraday's discovery that e.m.f. can be induced in a stationary coil by changing the magnetic flux linked with the coil. This principle is used in the transformer.

Consider the idealized arrangement of Fig. 6.14. Suppose a generator of negligible internal resistance is connected to a winding of N_1 turns. The p.d. at the generator terminals will cause a current to flow and this will set up a magnetic flux Φ. Suppose that this flux is constrained to follow the path indicated and thus

* See the author's *Applied Electromagnetism*, a companion volume in this series.

to link a second coil of N_2 turns. This constraint could be provided by an iron core and the effect of iron on magnetic fields will be discussed in the next chapter. In a real transformer not all the flux would link both coils, there would also be "leakage" flux but we are neglecting this effect. Let us also neglect the resistance of both windings. The arrangement of Fig. 6.14 with the various simplifying assumptions that have been mentioned is known as an "ideal" transformer and gives a good approximation to the behaviour of actual transformers in many cases.

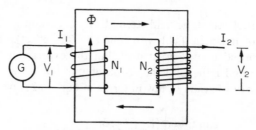

FIG. 6.14   *A simple transformer*

Since the winding connected to the generator has no resistance, the p.d. V_1 must be exactly equal to the e.m.f. induced in the winding (see p. 128). Thus

$$V_1 = \text{e.m.f.} = -N_1\frac{d\Phi_1}{dt} \tag{6.29}$$

Similarly

$$V_2 = -N_2\frac{d\Phi_2}{dt} \tag{6.30}$$

But

$$\Phi_1 = -\Phi_2 = \Phi$$

Thus

$$\frac{V_2}{V_1} = -\frac{N_2}{N_1} \tag{6.31}$$

Suppose now that the circuit of the second coil is closed, for example by connecting some resistance between its terminals. The p.d. V_2 will now circulate a current I_2. The m.m.f. of this current would tend to alter the magnetic flux, but the flux is fixed by its relationship to the generator p.d. V_1. Thus an extra current

I_1 has to flow in the winding connected to the generator to counterbalance the m.m.f. and we have

$$I_1 N_1 = I_2 N_2 \tag{6.32}$$

The total primary current is $I_1 + I_m$ where I_m, the magnetizing current which maintains the flux Φ, can be measured when the secondary winding is open-circuited. In dealing with ideal transformers it is usual to neglect the current I_m. In that case I_1 is the total primary current and we can write numerically

Turns ratio N_2/N_1 = secondary p.d./primary p.d.

= primary current/secondary current.

6.4. INDUCTANCE: A MECHANICAL EXPLANATION OF FARADAY'S LAW

We have found that electric charges experience forces when they travel across a magnetic field and also when they are at rest in a changing magnetic field. We could leave the matter there and ask no further questions about it. But readers of this book will by now be familiar with the author's intention to treat electrical phenomena as the mechanics of electric charge, just as ordinary mechanics deals with gravitating mass. So we ask ourselves what in ordinary mechanics corresponds to Faraday's law.

We notice first that Faraday's law deals with the effect of magnetic fields and that magnetic fields are always caused by moving electric charges. This had already been noticed when in the last chapter we were considering the direction of the forces between electric currents, because there we made use of the fact that magnetic energy is kinetic and that therefore the forces are such as to increase the magnetic field. This gives us the clue that we need. Since magnetic energy is kinetic, we can make the important deduction that electric charges possess inertia not only on account of their gravitating mass but also on account of their electrical charge. Moreover since the ratio of charge to mass in an electron is enormous (1.76×10^{11} coulomb/kg) it is not surprising that kinetic energy due to charge is very much bigger than kinetic energy due to mass, so that the latter can be ignored, except in special cases such as high-frequency oscillations in ionized gases.

The amount of the kinetic energy due to moving charges can be obtained from Faraday's law.

$$\text{Energy} = \int_0^t (\text{Volts} \times \text{Amps})\, dt$$

$$= \int_0^t \frac{d\Phi}{dt} I\, dt \qquad (6.33)$$

If the flux is proportional to the current, as it is in the absence of non-linear magnetic materials,

$$\Phi = LI \qquad (6.34)$$

where L is a constant.

Hence

$$\text{Energy} = \int_0^t LI \frac{dI}{dt}\, dt$$

$$= \tfrac{1}{2}LI^2 = \tfrac{1}{2}\Phi I \qquad (6.35)$$

Thus the kinetic energy can be expressed as the product of flux and current. If we compare the expressions for mechanical and electrical kinetic energy we have $\tfrac{1}{2}mv^2$ and $\tfrac{1}{2}LI^2$. Thus the equivalent expressions for momentum can be taken as mv and LI. Since $LI = \Phi$ we see that the magnetic flux is a measure of the electrokinetic momentum.* Thus Faraday's law is analogous to Newton's second law of motion: force is the rate of change of momentum.

Consider the simple example of Fig. 6.15. A battery of e.m.f. V and negligible internal resistance is connected through a switch to a circuit which has a small resistance R. When the switch is

---

* We have used the careful statement that Φ is a measure of momentum, because Φ does not have the dimensions of momentum. However, ΦI does have the dimensions of energy.

open the electric field along the wire is everywhere zero since no current can flow. When the switch is closed a current flows. What is the value of this current? Ohm's law suggests that the current will be large because the resistance is small. But there cannot be a large current all at once. The charges in the wire

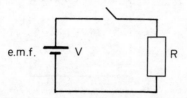

FIG. 6.15 *A battery connected to a resistance through a switch*

have to be accelerated and this will take time. Ohm's law by itself is not enough, we need to consider the kinetic energy associated with the current.

Allowance can be made for this kinetic energy by means of Faraday's law and we can write, using Kirchhoff's second law (p.75),

$$V - \frac{d\Phi}{dt} = RI \qquad (6.36)$$

where Φ is the flux linked with the circuit when the current in the circuit is I.

Equation (6.36) can be rewritten in terms of the constant L where $L = \Phi/I$ as

$$V = L\frac{dI}{dt} + RI \qquad (6.37)$$

and we note that this equation is analogous to the mechanical equation

$$F = m\frac{dv}{dt} + kv \qquad (6.38)$$

Thus $L(dI/dt)$ is a measure of the inertia force and RI of the fluid friction. Clearly the constant L is of great importance in the consideration of changing currents. It is called the *self-inductance* and depends on the geometry of the circuit. The reason for the name *self*-inductance is that the flux Φ in our example is due to the current in the circuit itself. In order to serve as a reminder, a

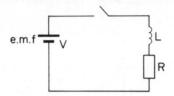

FIG. 6.16   *Inductance as a circuit element*

special symbol is used for self-inductance in drawing circuit diagrams, and this is shown in Fig. 6.16. The unit of inductance is the henry.

Equation 6.37 has the solution

$$I = \frac{V}{R}(1 - e^{-(R/L)t}) \qquad (6.39)$$

as may be shown by substitution or by reference to a book on differential equations. Figure 6.17 shows this relationship

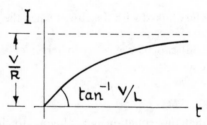

FIG. 6.17   *Build-up of current in the circuit*

graphically. Thus Ohm's law determines the final current, but the self-inductance L prevents an instantaneous rise to this final value. In fact the initial rate of rise of the current depends entirely on L. This is just what we know from experience of

mechanical systems. In a train, for instance, the initial acceleration depends on the inertia, whereas the final steady velocity
depends on the frictional forces. In fact Faraday's law has lost
much of its mystery.

Before leaving the discussion of inertia in current circuits we
must notice a very wonderful difference between mechanical and
electromagnetic inertia. Consider again the general behaviour of
a circuit as typified by eqn. (6.36). In this equation we identified
$d\Phi/dt$ with $L\,dI/dt$, because we assumed that the flux Φ was
caused by the current I in the circuit. However, it has already
been mentioned that charges experience forces in any changing
magnetic field and this field need not be caused by the motion of
the electric charges themselves. In the transformer of Fig. 6.14,
for instance, the charges in the secondary coil are acted upon by
the changing flux caused by a current in the primary coil or vice
versa. Thus the flux linked with a coil can be of two kinds: a self-
induced flux Φ_s and a mutual flux Φ_m provided by a current in
another coil.

Equation (6.36) can now be written

$$V_1 = \frac{d\Phi_s}{dt} + \frac{d\Phi_m}{dt} + RI_1 \qquad (6.40)$$

where the subscript 1 refers to the circuit under consideration.
Just as Φ_s can be written as LI_1, we can write Φ_m as MI_2, since it
is caused by the current I_2 in some other coil. M is called the
mutual inductance.

Then $$V_1 = L\frac{dI_1}{dt} + M\frac{dI_2}{dt} + RI_1 \qquad (6.41)$$

There is a force on the charges constituting current I_1 because
current I_2 is being changed. The mechanical analogy is that there
is a force on mass M_1 due to the acceleration of another mass M_2.
This would occur in a mechanical system in which the masses
were coupled together.

What is the coupling in the electrical case? Our discussion has shown that this coupling is provided by the magnetic flux and this is a remarkable discovery. Mechanical systems are coupled by material objects such as shafts, gears and belts. Electrical systems on the other hand do not require any such material connection. They do not even need wires or special dielectric or magnetic substances, all they need is a mutual flux. This makes it possible to transmit energy across empty spaces. The transformer is an application of this principle, for although we discussed in Fig. 6.14 a transformer with an iron core, there are many transformers which dispense with the iron. Another application of the same principle is given in induction motors, where it is possible to move the rotating part of the machine purely by changing the current in the stationary part. The change of momentum of the charges in the stator exerts a force on the charges in the rotor. There is no physical connection between stator and rotor. Energy and force are transmitted through empty space.

6.5. FURTHER DISCUSSION OF SELF AND MUTUAL INDUCTANCE

The chief reason for the interest which electrical engineers have in the idea of magnetic flux is that it provides a measure of the coupling of circuits. It is the flux linkage which matters. Where it is desirable to have a high electrical inertia and therefore a high flux linkage it is often convenient to wind a coil with many turns embracing the flux. By reference to Faraday's law, eqn. (6.26), it is clear that the self inductance for a coil of N_1 turns embracing a flux Φ is given by

$$L_1 = \frac{N_1 \Phi}{I_1} \tag{6.42}$$

and similarly the mutual inductance is given by

$$M_{12} = \frac{N_1 \Phi}{I_2} \tag{6.43}$$

Conversely when it is desired to reduce the electrical inertia, it is essential to reduce the flux linkage as far as possible.

In the expression for mutual inductance, eqn. (6.43), we have used the subscripts 1 and 2 to indicate the inductance of coil 1 due to current in coil 2. The question immediately arises, is there any relationship between M_{12} and M_{21}, where the latter expression defines the inductance of coil 2 due to current in coil 1? In the last chapter, Section 5.11, we derived the force on a coil in a magnetic field. If we use this expression for the force on coil 1, we have

$$F_1 = I_1 N_1 \frac{\partial \Phi}{\partial x}$$

$$= I_1 I_2 \frac{\partial M_{12}}{\partial x} \tag{6.44}$$

By Newton's third law of action and reaction

$$F_2 = -F_1 \tag{6.45}$$

where
$$F_2 = I_2 I_1 \frac{\partial M_{21}}{\partial x^1} \tag{6.46}$$

and
$$x^1 = -x$$

whence
$$\frac{\partial M_{12}}{\partial x} = \frac{\partial M_{21}}{\partial x} \tag{6.47}$$

and, since the mutual inductance is zero when the coils are far apart

$$M_{12} = M_{21} = M \tag{6.48}$$

This important result is very useful in cases where only one of the two quantities is easy to compute (see Ex. 8). The mutual inductance can be given either a positive or a negative sign. If the mutual flux adds the flux of self-inductance the sign of the mutual inductance is taken as positive, if the fluxes are in opposite directions the mutual inductance is negative.

We have assumed that the relationship between flux and current is one of simple proportionality, so that the inductances are

constant. In iron-cored circuits this is not true and the inductances vary with the current. Although the concept of inductance
in such non-linear cases is still a useful one, it is necessary in the
application of Faraday's law to go back to the basic concept of
magnetic flux in order to avoid mistakes.

SUMMARY

This chapter has been concerned with the interaction of electric
charges and magnetic fields.

It has been shown that an electric charge which is moving across
a magnetic field experiences a force. This motional force can be
used to produce electromotive forces in moving conductors. The
principle finds application in electric generators and motors in the
energy conversion from mechanical to electrical energy and vice
versa.

It has further been shown that an electric charge experiences a
force in a changing magnetic field and that this result and the
previous one are both covered by Faraday's law of electromagnetic
induction.

A mechanical explanation of this law has been given showing
that magnetic flux is a measure of the momentum of electric
charges and that the force arising from change of flux is analogous
to the mechanical force arising from change of momentum.

The action of a simple transformer has been discussed and the
terms self-inductance and mutual inductance have been defined
and elucidated.

NEW TERMS USED IN THIS CHAPTER

Term	Symbol	Unit (Abbreviation)	Definition
Self-Inductance	L	henry (H)	$L_1 = N_1\Phi/I_1$
Mutual Inductance	M	henry (H)	$M = N_1\Phi/I_2$
			$= N_2\Phi/I_1$
Primary magnetic constant	μ_0	henry/metre (H/m)	$B/H = \dfrac{\Phi/m^2}{I/m}$
			$= H/m$

Exercises

6.1. An electron is introduced with a finite velocity into an evacuated space where there is either (*a*) a uniform electric field, or (*b*) a uniform magnetic field. In each case the initial velocity of the electron is perpendicular to the direction of the field. Explain the shape of the path of the electron in each field.

6.2. Figure 6.18 shows a sectional view on a diametral plane of two concentric hollow metal spheres. A magnetic field *B* is applied perpendicularly to the plane of the figure and a p.d. of 50 V is maintained between the spheres, the inner one being positive with respect to the outer. The dotted line shows the path of an electron of energy 100 electron-volts.

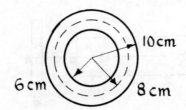

FIG. 6.18   *Electron moving in circular path*

If the magnetic field is reversed, estimate the p.d. which must be applied between the spheres in order to keep the electron moving in the same path.

(*Ans.* 164 V)

6.3. In a certain cathode-ray tube, focusing is achieved by means of a uniform axial magnetic field produced by a long solenoid. The flux density *B*, in T, is related to the current *I*, in amperes, by the relation $B = 5 \times 10^{-3}I$. It is found that a sharply-defined spot is obtained for successive values of the current of 0·9, 1·7, 2·6 and 3·5 A. Estimate the transit time between the anode hole and the screen.

(*Ans.* $8·16 \times 10^{-9}$ sec)

6.4. An electron-volt is the kinetic energy acquired by an electron falling through a p.d. of 1 volt.

An electron having an energy of 10^4 electron-volts is projected perpendicularly into a uniform magnetic field of strength 0·01 T. Determine the distance the electron will be from the point of entry after 3×10^{-9} sec.

(*Ans.* 33 mm)

6.5. A cathode-ray tube is surrounded by a coaxial solenoid producing a uniform magnetic field of 0·01 T. Electrons leave the small anode

aperture with an axial velocity of 6×10^7 m/sec and a maximum transverse velocity of 10^6 m/sec. The screen is 300 mm away from the anode. Estimate the size of the spot seen on the screen.
(*Ans.* 1·08 mm radius)

6.6. A stream of electrons, which are accelerated through 800 V, issues from an electron gun and is subjected to a uniform magnetic field of 56 mT perpendicular to it. If the stream is also acted upon by an electric field so that there is no deflection of the stream or change in the electron velocity, calculate the value of this electric field and show clearly its direction relative to the electron stream and the magnetic field.
(*Ans.* 944 kV/m)

6.7. Figure 6.19 shows an axially magnetized cylindrical magnet which can rotate on its axis. A conducting disc is mounted coaxially with the magnet and the disc can rotate independently of the magnet. A galvanometer is connected to the centre of the disc and to its circumference by means of a sliding contact.

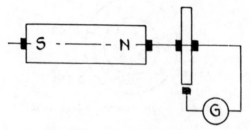

FIG. 6.19   *A homopolar generator*

Apply Faraday's law to decide whether the galvanometer will deflect when (*a*) the disc is rotated and the magnet is stationary, (*b*) the disc is stationary and the magnet rotates, and (*c*) both disc and magnet rotate in the same direction and at the same angular velocity.
(*Ans.* (*a*) Yes, (*b*) No, (*c*) Yes)

6.8. Define mutual inductance and find its dimensions.
A solenoid is uniformly wound with 180 turns of fine wire; the diameter of the solenoid is 6 cm and its length is 10 cm. A concentrated coil of 2 cm diameter, wound with 100 turns of fine wire, is placed coaxially and centrally within the solenoid. Estimate the mutual inductance of the coil in microhenries. Indicate how you would determine the flux linked with the solenoid due to a current in the small coil.
(*Ans.* ML^2/Q^2, 61)

6.9. In an experiment for the determination of the ohm by the Lorentz method, a circular copper disc is placed centrally inside a solenoid and spun about its axis which is coincident with the axis of the solenoid. The solenoid has a length of 30 cm and a mean diameter of 12 cm. It is uniformly wound with 4,000 turns. The diameter of the disc is 80 mm.

Calculate the e.m.f. induced between the centre and rim of the disc when it is rotated at 3,000 rev/min and the current in the solenoid is 5 A. (*Ans.* 19·5 mV)

6.10. A long metal tube of mean diameter 10 mm and thickness 0·25 mm is placed coaxially inside a long solenoid wound with 400 turns/mm. The resistivity of the metal of the cylinder is $5 \times 10^{-7}\Omega$m. If an alternating current of r.m.s. value 1 A at a frequency 100 kHz flows in the solenoid, calculate the power dissipated per metre length in the metal cylinder. (*Ans.* 5 W/m)

CHAPTER 7

The Magnetic Effects of Iron

7.1. THE USE OF MAGNETIC FLUX

We saw in the last chapter that moving electric charges have kinetic energy and momentum associated with them. This energy is specifically electromagnetic and is far bigger than the kinetic energy $\frac{1}{2}mv^2$ associated with the mass of the charges and their velocity. It is an energy which is described by the magnetic field surrounding electric currents and some writers say that the energy is stored in the magnetic field, although one could of course also say that the energy is stored in the currents. Since currents and magnetic fields always occur together, there is no possibility of deciding between the two ideas and either can be used as may be convenient. The energy belongs to the system.

Very many electrical devices make use of the forces that arise when the momentum of electric currents is changed. Faraday's law gives the magnitude of the resultant electromotive force. Moreover Faraday's law shows that the magnetic field can be used to couple circuits together and thus currents can be induced in circuits at a distance. This is such an extraordinarily useful effect that it is not surprising that electrical engineers spend much of their time trying to make good use of it and to produce strong magnetic fields. However, this is not easy because large currents tend to have large ohmic losses and energy is wasted in heating the coils, which probably will need to be cooled so that they do not overheat. One possible way in which the problem of losses may be overcome in the future is to use the property of super-conductivity which some materials have near the absolute zero of temperature. Super-conductors have no ohmic resistance and hence no losses. Much research effort is being spent in the quest for suitable super-conducting materials.

However, if the field strength required is not too large, there is happily another way in which magnetic fields can be produced without excessive ohmic loss, or with a small loss. Iron can be used, a material which has sources of magnetism which do not require energy to maintain them.

7.2. THE SOURCES OF MAGNETISM

In Chapter 5 we described Ampère's investigations into the relationship between electricity and magnetism. These investigations led Ampère to the knowledge that a small current loop and a small magnet have identical effects at a distance and he put forward the hypothesis that the action of magnets is due to *molecular currents*. This was a remarkable guess and one that was not at all obvious. Surely molecular currents that go on and on without batteries are a very queer phenomenon. Yet Ampère had hit on a truth which was unfolded by physicists about a hundred years later. Ampère's molecular currents are really atomic currents and they are of two kinds. First there is the current caused by orbital electrons moving around the nucleus and secondly there is the spin of the electrons about their own axes. Both these kinds of atomic current produce magnetic fields, nevertheless most materials have no magnetism because the atomic magnetic effects tend to cancel out when the atoms combine to form molecules. Three types of phenomena occur: diamagnetism, paramagnetism and ferromagnetism.

All materials have an inherent diamagnetic effect. If a material is put into a magnetic field the orbital electrons are acted upon by a force which changes the orbits in such a manner as to oppose the magnetic field. This is an example of Lenz's law. The effect is very slight and since it weakens the field it is generally the opposite of what the engineer wants.

Paramagnetism occurs in those substances in which the individual molecules possess a permanent magnetic dipole moment. When there is no external magnetic field the individual dipoles point in all directions, and the material has no resultant magnetic field. But when an external magnetic field is applied to paramagnetic substances the molecular dipole moments align themselves with the field and thus strengthen the field. In most paramagnetic substances the effect is slight although it is sufficient to cancel the diamagnetic effect, but there is one group of paramagnetic materials known as *ferromagnetics* in which the effect is enormous. Iron, cobalt and nickel are the chief ferromagnetic materials. The effect is strongest in iron and to the electrical engineer iron is a "magnetic" material, while all others are to all intents and purposes non-magnetic.

We can, therefore, say that there are two types of sources of magnetism. First there are the current sources, where magnetic fields surround electric currents. Secondly there are magnet sources, in which magnetic fields are produced by the aligned electron spins in iron.

7.3. HARD AND SOFT MAGNETIC MATERIALS

The magnetic moment of a spinning electron is constant, in fact the electron behaves as a tiny permanent magnet. Often it is desirable to have sizeable permanent magnets, containing immense numbers of electrons which would not be aligned under normal conditions. Much research effort by metallurgists has, however, produced alloys which are very nearly permanent magnets. These materials are extremely hard and brittle and once they have been magnetized, and the spins have been aligned to some extent, they are not greatly affected by external magnetic fields.

Permanent magnets are of course only useful if a constant

magnetic field is required. In alternating-current applications, for instance in transformers, we require a material which will strengthen the alternating magnetic field of the current. Such a material has to be magnetically soft. It is desirable that the electron spins should be aligned easily and that the direction of the alignment should be easily altered.

In actual practice there are no perfectly hard or soft magnetic materials, but the most useful materials are those that are either relatively hard or relatively soft. The actual behaviour can best be examined by the following experiment.

7.4. THE MAGNETIZATION CURVE

Figure 7.1 shows an iron ring on which a magnetizing winding has been wound which can be supplied with a current. There is also a secondary winding connected to a flux meter, which can measure the change of flux (the integral of voltage with time), when the magnetizing current is changed. We wish to observe the

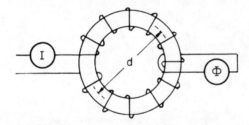

FIG. 7.1   *Ring used in the determination of a BH curve*

contribution of the iron to the magnetic field by observing the relationship between the magnetizing current and the flux. Better still we wish to obtain the relationship between the magnetic field strength H applied to the iron and the resultant flux density B in the iron. This we can do if H is uniform everywhere. Thus the magnetizing winding has to be wound uniformly and closely, and the width of the iron ring has to be small compared with its mean diameter. By the circuital law H will be given by

$$H = \frac{NI}{\pi d} \qquad (7.1)$$

where NI are the ampere-turns of the magnetizing winding and d is the mean diameter of the ring. The flux density will be given by

$$B = \frac{\Phi}{S} \qquad (7.2)$$

where Φ is the flux and S the cross-sectional area of the ring.

In carrying out the experiment it will be found that the change in B associated with a particular change in H depends not only on the values of H but also on the magnetic history of the iron. In order to obtain consistent results it is necessary to vary H

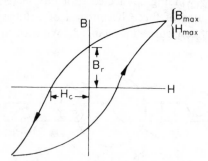

FIG. 7.2   *Cyclic magnetization curve*

several times between the same values of $\pm H_{max}$. The curve that will be obtained has the general shape of Fig. 7.2. It is symmetrical about its axes, but clearly there is a very complicated relationship between B and H.

The flux density B in Fig. 7.2 is not a single-valued function of H, but depends on the previous magnetic state of the iron. B tends to *lag* behind H; this phenomenon is called *hysteresis* and is associated with a loss of energy. Work has to be done to take the magnetization of the iron around the *hysteresis loop* of Fig. 7.2. When the applied magnetic field strength is zero there still remains a flux density of B_r. This is called the *remanence* of the iron. The value of H required to reduce the flux density to zero is called the *coercivity* and is marked on the curve as H_c. Large values of B_r

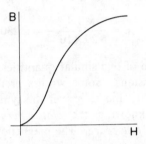

FIG. 7.3   *Saturation or reversal curve*

and especially of H_c are desirable in a permanent magnet. In fact the ideal permanent magnet would have infinite coercivity so that it would be impossible to demagnetize it. A soft magnetic material, on the other hand, should have a small coercivity and a narrow hysteresis loop.

In calculations on soft magnetic materials it is often possible to neglect the width of the hysteresis loop, although the energy loss due to hysteresis is still taken into account. In such cases engineers use the curve of Fig. 7.3, which is obtained by drawing a curve through the tips of various hysteresis loops for the material. Figure 7.3 is known as a magnetic reversal curve or a magnetic *saturation curve*. This name arises from the fact that, as H is increased, B reaches a limiting or *saturated* value. A typical value of flux density is about 1.4 T at a magnetic field strength of about 500 A/m.

A curve of B against H such as that of Fig. 7.3 is of great help because it enables the engineer to predict the flux which results from a certain applied magnetizing current. However, if we wish to know the contribution made to this flux by the iron, it is better to replot the x axis in terms of the flux density due to H in the absence of iron. This of course can be derived from the relationship $B_0 = \mu_0 H$, where $\mu_0 = 4\pi \times 10^{-7}$. Since μ_0 is a constant we need merely to change the scale of the axis. Thus for instance $H = 500$ A/m gives $B_0 - 6\cdot29 \times 10^{-4}$ T. The ratio of (flux density in iron/flux density in air) is called the *permeability* of the iron and is given the symbol μ_r. In the case under consideration

$$\mu_r = \frac{1\cdot4}{6\cdot29 \times 10^{-4}} = 2{,}230 \qquad (7.3)$$

Since μ_r is the ratio of two similar quantities it is a pure number and has no dimensions. Some writers prefer to call the ratio B/H the permeability and to give it the symbol μ. In that case $\mu = \mu_r \mu_0$ and μ_r is known as the *relative permeability*, while μ_0 is the *permeability of free space*. The magnetic susceptibility is defined in an analogous manner to electric susceptibility $\chi_m = \mu_r - 1$. The designer who wants to increase the magnetic flux is of course interested in the increase achieved by using a particular material

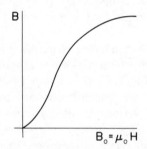

FIG. 7.4   *Curve to find relative permeability*

and so he is chiefly interested in the numerical value of permeability, i.e. in μ_r rather that $\mu_0 \mu_r$. In the special case of eqn. (7.3) it can be seen that the iron has increased the flux roughly 2,000 times. It is small wonder that iron is one of the most widely used materials in electrical engineering.

The permeability of iron is by no means constant. This can be deduced from Fig. 7.4. A typical curve of μ against H is shown in Fig. 7.5. A typical iron for a transformer curve has a permeability

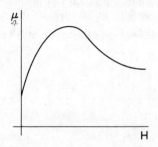

FIG. 7.5 *Permeability curve*

of a few thousand over most of the working range. Special alloys can be made to have a permeability as high as 500,000, but they are suitable for only limited application.

7.5. THE DOMAIN THEORY OF MAGNETISM

The shape of the curves in Figs. 7.2–7.5 is so curious that many readers of this book will want at least a few words of explanation. We saw in Section 7.2 that in paramagnetic substances there are atomic and molecular magnetic dipoles. This is true for iron, and the reason for the enormous magnetic effect in iron is that the dipoles tend to align themselves *without* the application of an external magnetic field. The alignment is complete throughout small volumes of the iron known as *domains*. These domains have linear dimensions of the order of 10^{-3} mm and are separated by domain walls in which the direction of magnetization changes from that of one domain to that of the adjacent one. The direction of each domain is along one of the easy directions of magnetization of a crystal parallel to the edges of the crystalline cubic lattice. In the bulk material the domains tend to cancel one another, but an apparently unmagnetized piece of iron consists of large numbers of very strongly magnetized domains. Consider now the effect on the domains when an external magnetic field is applied. In Fig. 7.6(a) four domains of equal size are shown schematically. When a small external field H_e is applied, the domain walls tend to move so as to increase those domains which are roughly in the

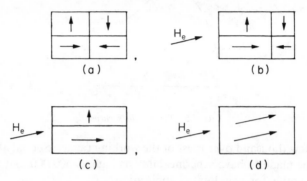

FIG. 7.6   *Magnetic domains*

direction of the applied field [Fig. 7.6(b)]. The magnetic field of the domains now begins to reinforce the applied magnetic field. This describes the foot of the curve in Figs. 7.3 and 7.4. If the external field is increased further, the opposing domains reverse their direction [Fig. 7.6(c)] and the magnetization curve rises very steeply. For larger values of H_e the direction of magnetization is forced from the easy crystalline direction to that of the applied field [Fig. 7.6(d)]. The iron is now saturated and the magnetization curve is almost flat.

7.6. MAGNETIC CIRCUITS

The thousandfold increase of flux which can be obtained by the use of iron is of tremendous help to the electrical engineer. There is also another related property of iron which is fully as important. Iron *guides* the flux and in a loose way of speaking an iron core can be said to *conduct* flux much as a copper conductor conducts current. Since the valuable thing about flux is its linkage with an electric circuit, it is clearly most important to be able to guide the flux in the right path.

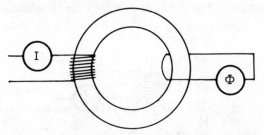

FIG. 7.7   *Ring with concentrated magnetizing winding*

Consider again the experiment of Fig. 7.1. The magnetizing winding was wound uniformly around the ring so that H and B should also be uniform. Now let the magnetizing winding be concentrated as in Fig. 7.7. If there are N turns in the coil, the magnetic field strength H due to the current in the winding alone is now about N times as strong near the magnetizing winding and very much weaker near the secondary winding. On the face of it one should expect that the flux-meter deflection for a certain change of magnetizing current would be greatly reduced. It is very puzzling to find that in fact the flux is apparently unaffected by the position of the magnetizing winding. If we try the experiment with a non-magnetic ring, the flux does depend dominantly on the position of the magnetizing winding, but when iron is used this position does not seem to matter. Now if the flux linking the secondary winding is unchanged, then the flux density is unchanged and a certain flux density B presupposes a certain magnetic field strength H in accordance with the magnetization curve. Since H due to the magnetizing coil is greatly reduced, there must be other sources of H to compensate for this reduction. Moreover, since this effect occurs only when there is a magnetic core it is clear that this core must somehow provide the additional magnetic field strength.

The understanding of this puzzle can be helped by considering the conduction of electric current around a circuit. We saw in Chapter 4 that the local electric field strength in the conductor is produced by local charges on the surface of the conductor. Consideration of Fig. 7.7 leads to a very similar conclusion. The

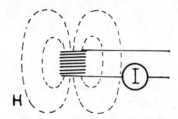

FIG. 7.8   *Field of the magnetizing coil in air*

field due to the magnetizing coil is symmetrical about the coil as shown in Fig. 7.8. When the coil is wound on a magnetic core, the domains of the core are acted upon by the field of the coil, and surface polarity appears on the core as indicated in Fig. 7.9. The magnetic field strength of any point is now the sum of H_1 due to surface polarity and H_2 due to current in the magnetizing coil.

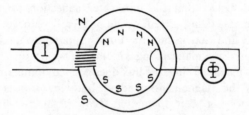

FIG. 7.9   *The effect of surface polarity*

H_1 describes a conservative field and H_2 the magnetomotive force of the winding. The experiment shows that $(H_1 + H_2)$ is nearly constant around the core. Thus H_1 and H_2 are opposed near the magnetizing winding and they add near the secondary winding.

The surface polarity ensures that most of the flux remains in the iron. Consider this with respect to Fig. 7.10. A flux Φ is directed towards a bend in an iron core. If the material were non-magnetic the direction of the flux would be unchanged. However, since it is iron, a small amount of flux emerging from the core sets up a strong surface polarity, and this turns the domains in the iron to follow the direction of the iron core.

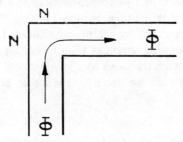

FIG. 7.10   *Direction of flux changed by surface polarity*

Any surface polarity implies of course a flux emerging from the iron. This is generally referred to as a leakage flux. The analogy of the *conduction* of flux and of electric charge here breaks down. Electric charges do not leak from a conductor when there is a surface charge, and in any case magnetic flux does not imply a flow of any substance, whereas current is a flow of charge. The exact analogy is between iron paths carrying magnetic flux and dielectric paths carrying electric flux. Because of the high permeability of iron the amount of leakage flux is generally small—perhaps a few per cent of the main flux. A dielectric circuit would have a far greater leakage flux because there is no electric phenomenon as strong as ferromagnetism.

The property of iron discussed in this section is very useful. The designer can use iron to "pipe" the magnetic flux to the place he wishes and he can put the magnetizing coil more or less anywhere on the iron core. Of course there is some loss of flux if the iron paths are long, but often this leakage flux does not matter greatly.

7.7. AN IRON RING WITH AN AIR GAP

In transformers it is possible to use iron cores which have no gaps in them. But in many electrical devices an air gap is essential so that there can be relative motion between the primary and the secondary winding. In rotating machines, for instance, the stationary and rotating windings both have iron cores, but there is an air gap between the cores to allow rotation. To understand the principles of such an arrangement let us examine the very

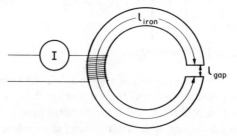

FIG. 7.11   *A ring with an air gap*

simple case shown in Fig. 7.11. Suppose that we wish to provide a magnetic flux density B in the gap, how many ampere-turns are required in the magnetizing winding?

It is clear that surface polarity will appear on the faces of the iron at the air gap and that the magnetic field in the gap will be similar to the electric field of a parallel-plate capacitor. If the surface area of the iron faces is large compared with the gap width, we can neglect the fringing flux (see Fig. 3.21, page 56) and consider B to be uniform in the gap. The poles on the surface of the iron are of course not free poles, such do not exist in nature, but the ends of the magnetic dipoles formed by domains in the iron. This means that tubes of magnetic flux are always continuous, whereas tubes of electric flux end on free charges. Thus the flux density in the iron is the same as the flux density in the

gap. However, the magnetic field strength is different. In the air
gap it is B/μ_0 and in the iron $B/\mu_r\mu_0$.

Now let us apply the circuital law of magnetism

$$\oint H\,dl = NI \qquad (7.4)$$

If we neglect leakage flux, B and H in the iron will be constant
around the ring. Thus

$$NI = \oint H\,dl = H_{\text{iron}}\,l_{\text{iron}} + H_{\text{gap}}\,l_{\text{gap}}$$

$$= \frac{Bl_{\text{iron}}}{\mu_r\mu_0} + \frac{Bl_{\text{gap}}}{\mu_0}$$

$$= \frac{Bl_{\text{gap}}}{\mu_0}\left(1 + \frac{l_{\text{iron}}}{\mu_r l_{\text{gap}}}\right) \qquad (7.5)$$

Suppose $\qquad \mu_r = 3{,}000 \quad$ and $\quad l_{\text{iron}} = 30 l_{\text{gap}}$

Then $\qquad NI = \dfrac{Bl_{\text{gap}}}{\mu_0}(1+0\cdot01) \qquad (7.6)$

This is a very interesting result. Only 1% of the magnetomotive
force is absorbed by the iron path, although it is so much longer;
99% of the m.m.f. is used in forcing the flux across the air gap.
Once again we see that it is easy to magnetize iron and difficult to
establish a magnetic field in air. This means that in designing
machines with air gaps, the designer, in order to save magnetizing
current, generally chooses a gap which is little more than a
mechanical clearance. The length of the iron path is not very
critical.

If the cross-sectional area of the ring is S, eqn. (7.5) can be
rewritten as

$$\text{m.m.f.} = \Phi\frac{l_{\text{gap}}}{S\mu_0}\left(1 + \frac{l_{\text{iron}}}{\mu_r l_{\text{gap}}}\right) \qquad (7.7)$$

By analogy with the electric circuit equation

$$\text{e.m.f.} = IR \qquad (7.8)$$

we can write

$$\text{m.m.f.} = \Phi\bar{R} \qquad (7.9)$$

and

$$\bar{R} = \frac{l_{gap}}{S\mu_0}\left(1 + \frac{l_{iron}}{\mu_r\, l_{gap}}\right) \qquad (7.10)$$

where $\bar{R}$ is the *reluctance* of the magnetic circuit.

The analogy is not complete because Φ and I are very different entities. Nothing flows when there is a magnetic flux. Moreover, Ohm's law tells us that R is independent of I, whereas $\bar{R}$ is not independent of Φ because the permeability varies with Φ.

However, eqn. (7.9) is useful in enabling us to draw an equivalent circuit.

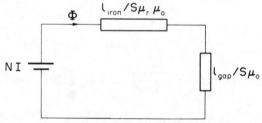

FIG. 7.12   *An equivalent magnetic circuit*

Such a circuit is illustrated in Fig. 7.12. It has to be used with care because, as we have seen, flux and current behave differently from one another, but readers of this book who like electrical circuits and distrust magnetism will use this device with good effect.

7.8. PERMANENT MAGNET CALCULATIONS

The magnetic field in the example of the last section was due to the current in the magnetizing winding. However, it is not necessary to have an external m.m.f. in iron which has very pronounced hysteresis, because in such a material the magnetic domains can themselves produce the magnetic field. Consider the magnet of Fig. 7.13, which illustrates both the H field and the B field. The

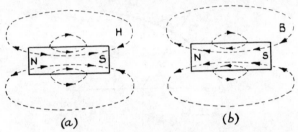

(a)   (b)

FIG. 7.13   *Field of a permanent magnet*

lines of B are continuous but the lines of H always run from north to south and are therefore discontinuous at the surface of the magnet. This reversal of H is of course implied by the statement that

$$\oint H\,dl = 0 \qquad (7.11)$$

Thus H and B are in opposite directions inside the magnet. This would be impossible in an ideally "soft" magnetic material in

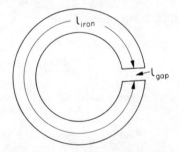

FIG. 7.14   *Permanent magnet with air gap*

which B and H are uniquely related by means of a curve such as shown in Fig. 7.3. It is, however, quite a possible situation if the relationship between B and H is that of Fig. 7.2, because there is a region of positive B and negative H. This is the region of interest in magnetic circuits which use permanent magnets.

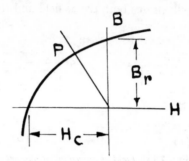

FIG. 7.15 *B–H curve of a permanent magnet*

Consider now a permanent magnet bent into a ring with an air gap as shown in Fig. 7.14 and let Fig. 7.15 be the relevant portion of the B–H curve. What is the flux density in the air gap?

The circuital law gives

$$\oint H \, dl = H_{\text{iron}} \, l_{\text{iron}} + H_{\text{gap}} \, l_{\text{gap}}$$

$$= 0 \tag{7.12}$$

Let the required flux density be B, then

$$B_{\text{iron}} = B_{\text{gap}} = B \tag{7.13}$$

and therefore

$$B_{\text{iron}} = \mu_0 H_{\text{gap}}$$

$$= -\frac{\mu_0 \, l_{\text{iron}}}{l_{\text{gap}}} H_{\text{iron}} \tag{7.14}$$

or

$$\frac{B_{\text{iron}}}{H_{\text{iron}}} = \frac{-\mu_0 \, l_{\text{iron}}}{l_{\text{gap}}} \tag{7.15}$$

This straight-line relationship has been plotted on Fig. 7.15. The point of intersection P gives the required flux density. Figure 7.15 illustrates why it is desirable to have a wide, gently sloping hysteresis curve. The actual value of the remanence B_r is less important, because the magnet does not operate at this point. In actual designs it is not usual to employ long permanent magnets of the shape shown in Fig. 7.14, because permanent magnets are expensive and so hard that they are impossible to machine except by grinding. Usually only a short part of the magnetic circuit is made of permanent magnet material. The rest is made of soft iron which is used to lead the flux to the required place.

7.9. MECHANICAL FORCE EXERTED BY A MAGNET

In Section 3.9 we considered the force per unit area on a charged conductor and found it to be

$$f = \tfrac{1}{2}ED \quad \text{N/m}^2 \tag{7.16}$$

where E and D are the electric field strength and electric flux density normal to the surface and just outside it. This result can be interpreted to mean that a tube of electric flux experiences a tension of $\tfrac{1}{2}ED$ per unit area. Since there is an exact correspondence between E and H, and D and B it follows that there is also a tension of $\tfrac{1}{2}HB$ per unit area in tubes of magnetic flux. This is a very general result and applies to all magnetic field problems. It is quite independent of the shape of the B–H curve. However, it does not give the complete answer because tubes of electric and magnetic flux also experience a pressure perpendicular to themselves. This pressure has the same value as the tension $\tfrac{1}{2}ED$ in the electric case and $\tfrac{1}{2}HB$ in the magnetic case. To obtain the resultant force on a body it is necessary to integrate the tensile stress and the pressure over the complete surface of the body. An exact treatment is beyond the scope of this book. We have, however, mentioned the tensile stress, because in many cases of symmetry there can be no sideways pressure, so that the tension is all that matters.

We have deduced that the magnetic tensile stress is given by

$$f = \tfrac{1}{2}HB \tag{7.17}$$

from considerations of the one-to-one correspondence between electric and magnetic quantities. Let us test the deduction by examining the force trying to close the air gap of the magnet in Fig. 7.14, which is shown to a larger scale in Fig. 7.16.

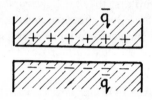

FIG. 7.16   *Surface polarity at the air gap*

Let the pole strength per unit area on the surfaces be $\pm \bar{q}$. Then the force per unit area is given by the average magnetic field strength at the surface multiplied by $\bar{q}$.

Thus $$f = \tfrac{1}{2}(H_{\text{gap}} - H_{\text{iron}})\bar{q} \qquad (7.18)$$

The negative sign is due to the reversal of direction of H. The discontinuity of H at the surface is due to the pole strength and, by Gauss's theorem, the contribution to the magnetic field

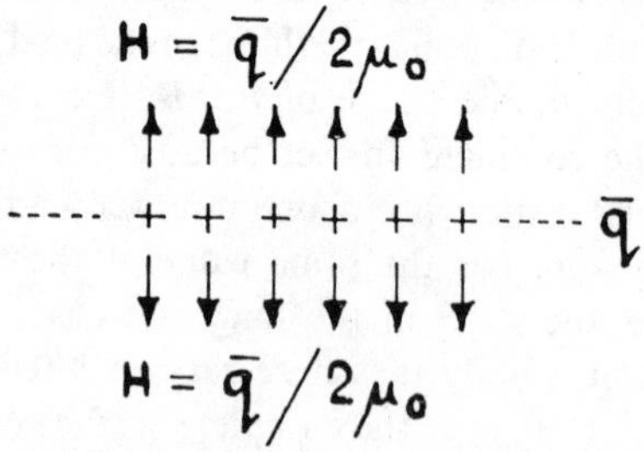

$$\mathsf{H} = \bar{q}\big/2\mu_0$$

$$\uparrow \ \uparrow \ \uparrow \ \uparrow \ \uparrow \ \uparrow$$

$$\text{-----} + + + + + + \text{-----} \ \bar{q}$$

$$\downarrow \ \downarrow \ \downarrow \ \downarrow \ \downarrow \ \downarrow$$

$$\mathsf{H} = \bar{q}\big/2\mu_0$$

FIG. 7.17   *Magnetic field of surface polarity*

strength made by $\bar{q}$ is $\bar{q}/2\mu_0$ in either direction as shown in Fig. 7.17. Thus the discontinuity in H is $\bar{q}/\mu_0$. Thus

$$H_{\text{gap}} - \frac{\bar{q}}{\mu_0} = -H_{\text{iron}} \qquad (7.19)$$

Also from eqn. (7.12)

$$H_{\text{gap}} = -\frac{l_{\text{iron}}}{l_{\text{gap}}} H_{\text{iron}} \tag{7.20}$$

Thus
$$f = \tfrac{1}{2}(H_{\text{gap}} - H_{\text{iron}})\mu_0 (H_{\text{gap}} + H_{\text{iron}})$$

$$= \tfrac{1}{2}H_{\text{gap}} B_{\text{gap}} \left(1 - \left\{\frac{l_{\text{gap}}}{l_{\text{iron}}}\right\}^2\right) \tag{7.21}$$

Apparently eqn. (7.17) is at variance with eqn. (7.21). However, this is not so. The trouble is that Fig. 7.16 shows only part of the iron. Equation (7.17) correctly gives the total force which tries to close the gap. Equation (7.21) gives that part of the force which acts on the iron surface in the gap, but leaves out the force associated with the leakage flux. Happily there will be very little difference in the numerical answer obtained with either formula. Happier still it is the simpler formula which is the correct one.

The principle explained in this section leads to many useful applications in engineering practice, perhaps the most obvious one being that of a lifting magnet. In such magnets a coil surrounds a soft iron core. When current flows in the coil the core is magnetized and will lift other magnetic materials, which can be dropped again conveniently by switching off the current.

7.10. LOSSES IN IRON SUBJECTED TO ALTERNATING MAGNETIC FIELDS

7.10.1. *Hysteresis Loss*

Work has to be done to take the magnetic state of the material around the hysteresis loop typified by Fig. 7.2. Thus whenever iron is used in alternating current applications such as transformers, there is a hysteresis loss.

Consider the ring specimen of Fig. 7.1. Let the magnetizing

current I be supplied by a generator and let it be changed from $+I_{max}$ to $-I_{max}$ and back again to $+I_{max}$, and let this process take time τ.

Then the work done by the generator will be

$$W = \int_0^\tau -eI\,\mathrm{d}t \qquad (7.22)$$

where e is the e.m.f. induced by the change of flux in the iron.

Thus
$$e = -N\frac{\mathrm{d}\Phi}{\mathrm{d}t}$$

$$= -NS\frac{\mathrm{d}B}{\mathrm{d}t} \qquad (7.23)$$

where S is the cross-sectional area of the ring and N the number of turns of the winding. Also the circuital law gives

$$\oint H\,\mathrm{d}l = Hl = NI \qquad (7.24)$$

whence
$$W = \int_0^\tau SlH\frac{\mathrm{d}B}{\mathrm{d}t}\,\mathrm{d}t$$

$$= Sl\oint H\,\mathrm{d}B \qquad (7.25)$$

where the integral is taken around the complete magnetic cycle. This integral is the area of the hysteresis loop. Thus eqn. (7.25) states that the hysteresis loss per cycle is given by the volume of iron multiplied by the area of the hysteresis loop. Energy is supplied by the source of the current and reappears as heat in the iron.

The energy loss per cycle is constant, so the total loss depends on the number of cycles. The loss due to hysteresis is therefore proportional to frequency. It is also important to know how the loss varies if the maximum flux density is changed. Unfortunately there is no mathematical equation which fits the hysteresis loop, because the loop depends on the very complicated arrangement of large numbers of magnetic domains. Thus recourse is often had to an empirical relation. From tests it is found that the area of many hysteresis loops is proportional to $(B_{max})^x$, where x has a value in the range 1·5–2·3. Thus the power loss can be written as

$$P_h = kfB^x \qquad (7.26)$$

where k is a constant and f is the frequency.

7.10.2. Eddy Current Loss

Since iron is a conductor, there will be currents induced in it, when the flux is changed. These currents are called eddy currents because they circulate in the iron. Because of ohmic resistance, eddy currents cause a loss of energy. Hysteresis and eddy current loss form the two components of iron loss.

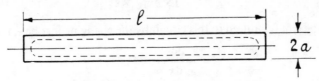

FIG. 7.18 *Eddy current path in an iron lamination*

Consider a rectangular block of iron as shown in Fig. 7.18. This block can be thought of as being part of one of the iron laminations which are used for the cores of transformers and rotating electrical machines. The laminations are placed parallel to the magnetic field. Let there be an alternating flux density perpendicular to the plane of the paper in Fig. 7.18 and let it be given by

$$B = B_m \sin \omega t \qquad (7.27)$$

The dotted line shows a typical path for the eddy current. Let us redraw the figure to a bigger scale (Fig. 7.19) and neglect the curved ends of the eddy current paths. Consider the current

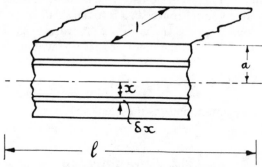

FIG. 7.19   *Calculation of eddy current loss*

flowing in two thin flat strips of width δx and depth unity. Let the resistivity of the material be ρ. Then the resistance of the circuit formed by the two thin strips is

$$R = \rho \frac{2l}{1 \times \delta x} = \frac{2\rho l}{\delta x} \qquad (7.28)$$

Let the e.m.f. induced in this circuit by the changing flux be e, then

$$e = -\frac{d\Phi}{dx}$$

$$= -2xl\omega B_m \cos \omega t \qquad (7.29)$$

and the ohmic power loss is

$$\frac{e^2}{R} = \frac{2x^2 l\omega^2 \, \delta x B_m^2 \cos^2 \omega t}{\rho} \qquad (7.30)$$

Thus the mean power loss is

$$\delta P = \frac{l\omega^2 B_m^2 x^2 \, \delta x}{\rho} \qquad (7.31)$$

and for the whole lamination

$$P = \int_0^a l\omega^2 B_m^2 x^2 \, dx$$

$$= \frac{l\omega^2 B_m^2 a^3}{3\rho} \tag{7.32}$$

The volume of the lamination is $2al$. Thus the power lost per unit volume is given by

$$P_e = \frac{\omega^2 B_m^2 a^2}{6\rho}$$

$$= \frac{2\pi^2 f^2 B_m^2 a^2}{3\rho} \tag{7.33}$$

Thus the eddy current loss per unit volume varies as the square of the frequency and the square of the lamination thickness. It varies inversely as the resistivity of the iron. In order to reduce the loss, designers use an alloy of high resistivity. The laminations are also made as thin as convenient. A thickness of $\frac{1}{2}$ mm is common.

Although eqn. (7.33) gives a good idea of the variation of eddy current loss with various parameters, it does not give accurate answers. Consider the assumptions we have made. We have assumed that the flux density is not affected by the eddy current. This is only true if the laminations are thin enough and the frequency is low. The assumption is correct for transformers and machines at power frequencies. At higher frequencies it is necessary to subdivide the iron more finely in order to reduce eddy currents. Iron dust can then be used. At radio frequencies it may be necessary to use different materials altogether. These are the ferrites, non-metallic magnetic materials which are insulators and have very high resistivity.

Another assumption underlying eqn. (7.33) is that the permeability of the iron is constant throughout the cycle. This is not true. If the formula is applied to eddy currents in homogeneous metals like copper or aluminium, it gives the right answer for the

loss due to eddy currents at low frequencies. But iron is made up
of domains and is not a homogeneous substance. It is found in
practice that the eddy current loss in iron is considerably higher
than the loss predicted by eqn. (7.33).

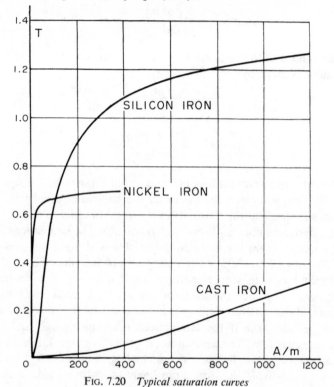

FIG. 7.20   *Typical saturation curves*

Eddy currents are not always undesirable. Some devices such
as eddy current couplings depend on these currents and in induction furnaces it is the heat generated by the eddy currents which
causes the metal to melt. A designer of an induction furnace tries
to design for maximum eddy currents loss. Hysteresis effect also
can sometimes be put to good use, as for instance in small
electric motors.

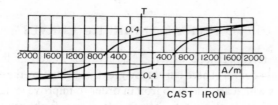

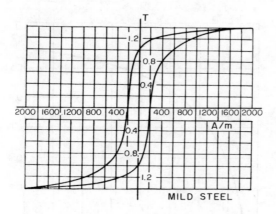

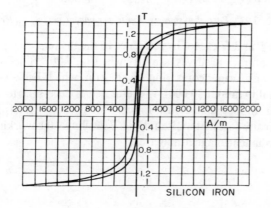

FIG. 7.21   *Typical hysteresis loops*

7.11. TYPICAL MAGNETIC CURVES

Every ferromagnetic material has its own hysteresis loop and saturation curve. These curves are greatly affected by the alloying elements which are added to iron in order to improve its magnetic properties. It is, therefore, not possible to specify the relationship

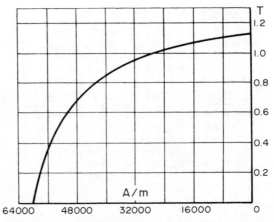

FIG. 7.22   *Part of hysteresis loop of permanent magnet material*

between B and H except by reference to the makers' curves. Some typical saturation curves are given in Fig. 7.20. Silicon iron is used in transformer cores, and nickel iron, which has a very high permeability (about 80,000), is useful if it is necessary to screen a region from magnetic fields. Figure 7.21 gives hysteresis loops of magnetically soft materials and Fig. 7.22 shows the working range of a permanent magnet material.

SUMMARY

The magnetic effects of iron are twofold. First, iron provides sources of magnetism additional to those provided by electric currents. Soft iron can be used as a core of a coil and, if there is no air gap, the flux will be increased by the order of some thousands. Hard permanent magnets can be used to provide a magnetic field without the help of electric currents.

Secondly, iron acts as a guide or conductor of magnetic fields. This enables the designer to use magnetic circuits.

The action of iron can be explained by the domain theory of ferromagnetism.

In alternating magnetic fields iron experiences hysteresis and eddy current losses.

NEW TERMS USED IN THIS CHAPTER

Term	Symbol	Unit (Abbreviation)	Definition
Permeability (relative)	μ_r	pure number	$\mu_r = B_{\text{iron}}/B_{\text{free space}}$
Permeability (absolute)	μ	henry/metre (H/m)	$\mu = \mu_r \mu_0$
Magnetic Susceptibility	χ_m	pure number	$\chi_m = \mu_r - 1$
Reluctance	$\overline{R}$	amperes/weber (A/Wb)	$\overline{R} = \text{m.m.f.}/\Phi$
Remanence	B_r	tesla (T)	B at zero H
Coercivity	H_c	amperes/metre (A/m)	H at zero B

Exercises

7.1. The magnetic energy associated with a current is given by $\frac{1}{2}LI^2$. Use
this expression to find the energy stored in the magnetic field of the ring
of Fig. 7.1 in terms of H and B. Show how your result can be applied
to magnetic fields of arbitrary shape. (*Hint:* Divide the field into tubes
of flux).
(*Ans.* $\frac{1}{2}HB$ per unit volume)

7.2. An iron core 180 cm long and 60 cm² in cross-section, magnetized
uniformly by an alternating current at a frequency of 50 Hz, had a total
loss of 900 watts. The hysteresis loss under the conditions of test was
370 J/m³ per cycle.
      Estimate the total loss of the core at the same maximum flux density
if the frequency is reduced to 40 Hz.
(*Ans.* 609 watt)

7.3. A magnetic circuit consists of an annular ring of iron of mean diameter
15 cm with a single narrow saw-cut through the ring, the plane of the
cut containing the axis of the ring. The width of the cut is 1 mm and
the cross-section of the ring is 2×2 cm. The iron of the ring has a
reversal curve given by the following table:

H	A/m	40	80	120	160	800	1600	3200
B	T	0·37	0·72	0·92	1·04	1·40	1·47	1·55

A coil of 400 turns is uniformly wound on the ring. Estimate the current
in the coil required to cause a flux of 0·28 mWb in the air gap formed
by the saw-cut.
      Explain in general terms why the simple method of estimation suitable
for the above conditions may give an appreciable error if the flux in
the air gap is to be about 0·6 mWb or larger.
(*Ans.* 1·49 A)

7.4. Explain the term *permeability*.
      A steel core is built of stampings, the relevant dimensions of which
are given in centimetres in Fig. 7.23. The stampings are assembled so
that there is an air gap 2 mm long between the central limb and the
base. The two outer limbs are wound with equal coils connected in
series. Part of the saturation curve for the steel is given by the following
table:

H	A/m	560	800	1040
B	T	1·36	1·41	1·438

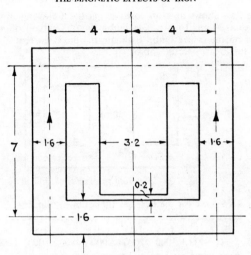

FIG. 7.23   *Relating to exercise* 7.4

Neglecting effects of leakage and fringing, estimate the value of the flux density in the iron when the coil on each outer limb carries 2400 ampere-turns and has a polarity such that it will cause an m.m.f. in its own limb in the direction shown in Fig. 7.23. If the polarity of one coil is reversed estimate the ampere-turns required on the outer limbs in order that the magnitude of the flux density in these limbs shall be unaltered.

(*Ans.* 1·41 T, 238 ampere-turns)

7.5. A magnetic circuit consists of a ring of steel of uniform cross-section and a mean perimeter of 450 mm with a uniform air gap of length 0·5 mm in series with the steel. The circuit is wound with a coil of 500 turns. Part of the hysteresis loop for the steel when it is magnetized to a maximum flux density of 1·1 T is given by the following table:

H	A/m	560	320	160	0	− 80	− 160
B	T	1·1	1·07	1·02	0·9	0·7	0

Allowing for fringing flux by taking the effective cross-section of the gap to be 5% larger than that of the steel, estimate

(1) the current which will bring the mean value of the flux density in the steel to 1·1 T and

(2) the flux density in the steel if the current is reduced from that value to 0·5 amps.

(*Ans.* 1·34 A, 0·74 T)

7.6. Figure 7.24 shows an iron core having the following cross-sectional area: centre limb 16cm², side limbs 10cm². A coil of 450 turns is wound on the centre limb. Estimate the direct current in the coil to give a flux density in the gap of 1·1 T.

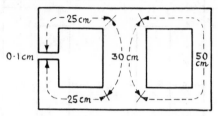

FIG. 7.24   *Relating to exercise 7.6*

The *B–H* curve for the iron has the following values

B	1	1·35	1·45	1·5	1·54	1·58	1·6
H	400	1000	2000	3000	4000	5000	6000

(*Ans.* 6·5 A)

7.7. An iron core 90mm long and 1000mm² in cross-section was magnetized uniformly by a winding carrying an alternating current. The following measurements were made:

Frequency (Hz)	Max. flux density (T)	Power loss in core (watts)
40	0·5	8
60	0·9	40

Estimate the area of the hysteresis loop for the test at 40 Hz assuming that this area varies as $B^{1·7}$.
(*Ans.* 1,520)

7.8. A small transformer is to have an iron core of uniform cross-section 1000 mm², and total volume 800 cm³. When supplied from 200 V, 50 Hz mains the peak flux density is to be $B = 0·9$ T² for which $H = 240$ A/m.
       Calculate the number of turns on the primary winding and the peak value of the magnetizing current.
       (*Ans.* 1,000, 192mA)

7.9. A horseshoe magnet has a cross-sectional area of 1000 mm² at each pole face. It is to be designed to lift an iron plate of mass 100kg. What is the minimum flux density required at the pole faces?
       (*Ans.* 1·11 T)

7.10. A permanent magnet is to be designed to produce a flux density B in an air gap of cross-sectional area S and length g. What point on the magnetization curve is the best operating point, if the volume of the permanent magnet is to be a minimum?

(*Ans.* The point that makes the product $-HB$ a maximum.)

CHAPTER 8

Electromagnetic Radiation

8.1. MAXWELL'S EQUATIONS

In Chapter 5 we discussed the magnetic effect of electric currents. We made use of the equivalence of current loops and magnetic dipoles to develop the circuital law $\oint H \, dl = I$ and this law has been extremely useful to us. It is in fact one of the most widely used relationships in electromagnetism.

However, sometimes the circuital law is unsatisfactory and at other times it is plainly wrong. It is wrong for instance in all problems involving radio waves. We know by experience that magnetic fields can be transmitted across empty space in which there are no electric currents. Surely it cannot be true for radio waves that $\oint H \, dl = 0$, since this would mean that the magnetic field was conservative and could not transmit energy.

An example in which the circuital law is unsatisfactory is given by the simple circuit of Fig. 8.1, in which an a.c. generator is connected to a capacitor through a resistor. We wish to investigate the magnetic field due to the alternating current. There is no difficulty in the application of the circuital law except at the capacitor. If the circuit embraces the wire or the resistor $\oint H \, dl = I$, but if the circuit integral is taken around the empty space between the capacitor plates it looks as though we should have to write $\oint H \, dl = 0$ because there is no current between the plates. This would imply a discontinuity in the magnetic field which seems most unlikely.

A difficulty of this kind should not cause dismay. Clearly we are on the verge of an important discovery which will increase our understanding of electromagnetism. Alternatively there may of course be a mistake in the derivation of the circuital law. If the reader will check this derivation, he should satisfy himself that there was no mistake. However, a careful reading will show that there was an assumption, namely that the current was steady and continuous, whereas in Fig. 8.1 we are dealing with an alternating current.

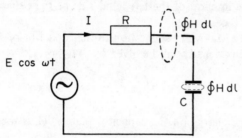

FIG. 8.1   *Illustrating a difficulty in applying the magnetic circuital law*

The solution of the difficulty is due to Maxwell, who built upon Faraday's ideas. It will be remembered that Faraday's great contribution was the idea that a changing magnetic flux causes an e.m.f. Maxwell put forward the complementary idea that a changing electric flux causes an m.m.f.

An electric charge is a centre of electric flux and hence a motion of electric charges, which constitutes a current, can be thought of as carrying an electric flux with it. But an electric flux need not be due to *local* charges. Distant charges may be the cause of this flux. Thus in order to account for both types of flux-change Maxwell proposed the law

$$\oint H\,dl = I + \frac{d\Psi}{dt} \tag{8.1}$$

where I is the current traversing the loop and Ψ is the electric flux from other charges. The flux can be written as

$$\Psi = \iint_S D_n\,dS \tag{8.2}$$

If the circuit is stationary

$$\frac{d}{dt}\iint_S D_n \, dS = \iint_S \frac{dD_n}{dt} \, dS \tag{8.3}$$

and then

$$\oint H \, dl = I + \iint_S \frac{dD_n}{dt} \, dS \tag{8.4}$$

where the line integral of the left hand side of the equation embraces the surface S.

Let us apply Maxwell's suggestion to the capacitor of Fig. 8.1. The electric flux density will be given by

$$D_n = \frac{Q}{S} \tag{8.5}$$

where Q is the charge on the capacitor plates. Thus the electric flux is

$$\Psi = Q \tag{8.6}$$

Also

$$I = \frac{dQ}{dt} \tag{8.7}$$

and therefore

$$I = \frac{d\Psi}{dt} \tag{8.8}$$

Thus the circuital law gives

$$\oint H \, dl = I + 0 = I \tag{8.9}$$

where it links the current, and

$$\oint H \, dl = 0 + \frac{d\Psi}{dt} = 0 + I = I \tag{8.10}$$

where it links the space between the capacitor plates. There is no discontinuity of H and the difficulty is resolved. The apparent discontinuity of H arises because the current is apparently discontinuous at the plates. But a current cannot just appear or disappear. If it comes to a break in the circuit then there will be charges near that break and these charges satisfy the continuity eqn. (8.7). Maxwell's m.m.f. eqn. (8.1) makes the magnetic field continuous by taking into account the continuity of the flow of electric charges. When we were discussing steady currents, this continuity was already inherent in the currents themselves, so there was no need for the additional term introduced by Maxwell. However, when we are dealing with changing currents we need the complete expression of eqn. (8.1).

Maxwell's expression marks a tremendous advance in electromagnetism. We shall devote the rest of this chapter to the exploration of some of the consequences of his idea. At this stage it is worth noting that eqn. (8.1) enables us to work out the m.m.f. due to currents and charges in open-ended conductors such as radio aerials and also the m.m.f. in empty space, which is associated with radio waves.

When there are no currents, Maxwell's statement and Faraday's statement can be written

$$\oint H \, dl = \frac{d\Psi}{dt} \qquad (8.11)$$

and

$$\oint E \, dl = -\frac{d\Phi}{dt} \qquad (8.12)$$

The two statements are generally called Maxwell's equations.

8.2. MAGNETOMOTIVE FORCE AROUND A CAPACITOR CONTAINING A DIELECTRIC MATERIAL

In Section 3.7 we discussed the action of a polarizable insulating material between the plates of a capacitor. In such a material the electric charges are fixed to the atomic lattice, but when an electric field is applied, the material can be polarized and there is a net transference of charge from one side of the material to the other.

The amount of charge displacement q is related to the applied field by the expression

$$q = S\chi_e \, \varepsilon_0 \, E \qquad (8.13)$$

where S is the area of each capacitor plate and χ_e is the electric susceptibility. The charge on the capacitor plates is related by the expression

$$Q = S\varepsilon_r \varepsilon_0 \, E \qquad (8.14)$$

where ε_r is the permittivity of the material.

Also
$$S\varepsilon_0 \, E = Q - q \qquad (8.15)$$

and hence
$$\varepsilon_r = 1 + \chi_e \qquad (8.16)$$

An outside observer cannot distinguish between electrons that are bound to the lattice and free electrons. To such an observer any transference of charge is an electric current and the transference of q is therefore called the polarization current. We thus have a polarization current of

$$I_p = \frac{dq}{dt} \qquad (8.17)$$

and the m.m.f. due to this current is

$$\oint H_p \, dl = I_p = \frac{dq}{dt} \qquad (8.18)$$

Thus the m.m.f. due to polarization is

$$\oint H_p \, dl = \frac{\chi_e}{\varepsilon_r} \frac{dQ}{dt}$$

$$= \frac{\chi_e}{\varepsilon_r} \frac{d\Psi}{dt} \qquad (8.19)$$

The polarization current is, therefore, not sufficient to give the correct total m.m.f. demanded by equation (8.11), and clearly it cannot be since q is smaller than Q. The missing part is

$$\frac{d\Psi}{dt} - \frac{\chi_e}{\varepsilon_r}\frac{d\Psi}{dt} = \frac{\varepsilon_r - \chi_e}{\varepsilon_r}\frac{d\Psi}{dt}$$

$$= \frac{1}{\varepsilon_r}\frac{d\Psi}{dt} \qquad (8.20)$$

This can be written in terms of the electric field strength

$$\frac{1}{\varepsilon_r}\frac{d\Psi}{dt} = S\varepsilon_0\frac{dE}{dt} \qquad (8.21)$$

Thus we must always add to the m.m.f. the change of electric flux associated with the changing electric field strength, quite apart from any sort of current whether it is conduction current, polarization current or convection current.

Some writers insist that *all* m.m.f. must be due to some form of current. In that case $S\varepsilon_0(dE/dt)$ must be some sort of a current and it is given the name *displacement* current. Other writers prefer to think of $S\varepsilon_0(dE/dt)$ as being the polarization current of free space. Neither idea is very helpful. It is preferable to make a clear distinction between the transference of electric charge, which constitutes current, and the change of electric flux in free space. Both effects have to be added to obtain the m.m.f.

8.3. ELECTROMAGNETIC WAVES

The last section showed that on Maxwell's hypothesis we must always introduce an additional term into the m.m.f. equation, because the m.m.f. is equal to the total rate of change of electric flux. As a consequence of Maxwell's modification the m.m.f. equation is now applicable to open-ended wires as well as to closed circuits. This is satisfactory, but it is only a very small part of the story.

The most startling consequence of Maxwell's idea has still to be shown. Consider the electromagnetic relationships in free space. Before Maxwell these were

$$\oint H \, \mathrm{d}l = 0 \tag{8.22}$$

and
$$\oint E \, \mathrm{d}l = -\frac{\mathrm{d}\Phi}{\mathrm{d}t} \tag{8.12}$$

but after Maxwell

$$\oint H \, \mathrm{d}l = \frac{\mathrm{d}\Psi}{\mathrm{d}t} \tag{8.11}$$

and
$$\oint E \, \mathrm{d}l = -\frac{\mathrm{d}\Phi}{\mathrm{d}t} \tag{8.12}$$

Let us examine the simplest possible case of an electromagnetic field. Let the electric field strength be entirely in the x direction

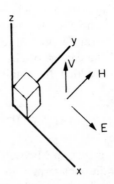

FIG. 8.2   *A plane electromagnetic field*

and the magnetic field strength entirely in the y direction and let these field strengths be the same throughout a plane parallel to the xy plane (Fig. 8.2). Such a configuration could arise from a large *current sheet*, in which the current flows in the xy plane and along the direction parallel to the x axis.

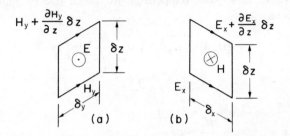

FIG. 8.3   *Relationship between the electric and magnetic field*

Let us apply the electromagnetic relationships to two small rectangles as indicated in Fig. 8.3 by integrating in an anti-clockwise direction around the rectangles. Without Maxwell's term the m.m.f. equation is given by

$$\oint H \, dl = H_y \delta y - \left(H_y + \frac{\partial H_y}{\partial z} \delta z \right) \delta y$$

$$= 0 \tag{8.23}$$

whence
$$\frac{\partial H_y}{\partial z} = 0 \tag{8.24}$$

This means that H_y does not vary in the z direction, and since we have already postulated that it does not vary in the x and y directions, it appears that H varies only with time.

Now the magnetic field is due to a current; at the surface of the current sheet it will vary in time in the same manner as the current varies, and thus the magnetic field will be in-phase with the current. But this means that the magnetic field *everywhere* will be in-phase with the current, because we have shown the field to be constant throughout space. This is unbelievable, because it means that the distant magnetic field changes instantaneously with the current, without any message having reached it to give information about the change of current. Clearly there is something wrong.

Consider now what happens if we make use of Maxwell's equations

$$\oint H\,dl = H_y\,\delta_y - \left(H_y + \frac{\partial H_y}{\partial z}\delta z\right)\delta y$$

$$= \varepsilon_0\,\delta z\,\delta y\,\frac{\partial E_x}{\partial t} \tag{8.25}$$

whence

$$-\frac{\partial H_y}{\partial z} = \varepsilon_0\frac{\partial E_x}{\partial t} \tag{8.26}$$

Similarly

$$\oint E\,dl = E_x\,\delta x - \left(E_x + \frac{\partial E_x}{\partial z}\delta z\right)\delta x$$

$$= \mu_0\,\delta z\,\delta x\,\frac{\partial H_y}{\partial t} \tag{8.27}$$

whence

$$-\frac{\partial E_x}{\partial z} = \mu_0\frac{\partial H_y}{\partial t} \tag{8.28}$$

If we eliminate either E or H, eqn. (8.26) and (8.28) give

$$\frac{\partial^2 H_y}{\partial z^2} = \mu_0\varepsilon_0\frac{\partial^2 H_y}{\partial t^2} \tag{8.29}$$

and

$$\frac{\partial^2 E_x}{\partial z^2} = \mu_0\varepsilon_0\frac{\partial^2 E_x}{\partial t^2} \tag{8.30}$$

Equations of this type often appear in engineering problems. They have solutions of the form

$$H_y = F(z \pm vt) \tag{8.31}$$

and

$$E_x = f(z \pm vt) \tag{8.32}$$

where F and f are any arbitrary functions. The correctness of eqn. (8.31) and (8.32) can be tested by substitution into eqn. (8.29) and (8.30).

The form of the solution shows that the values of H and E are unchanged, if $(z \pm vt)$ is unchanged and hence

if $$z \pm vt = \text{constant}$$

or if $$\frac{dz}{dt} = \mp v \qquad (8.33)$$

The equations, therefore, describe waves of constant magnitude travelling down or up the axis of z with constant velocity v. The equations are known as *wave equations*. Thus Maxwell's hypothesis has transformed the solution of the problem. In particular, the magnetic field is not constant throughout all space but consists of waves travelling in a direction perpendicular to the current sheet.

The velocity of the waves is given by v, where

$$v^2 = \frac{1}{\mu_0 \varepsilon_0} \qquad (8.34)$$

Now $\mu_0 = 4\pi \times 10^{-7}$ and $\varepsilon_0 = 8 \cdot 854 \times 10^{-12}$. This gives

$$v = 2 \cdot 998 \times 10^8 \text{m/s} \qquad (8.35)$$

But this figure was known to Maxwell as the velocity of light, which had been calculated long before by astronomers. Maxwell made the momentous deduction that electromagnetic waves travel with the speed of light and that light is an electromagnetic phenomenon. Many investigators before Maxwell had felt intuitively that electromagnetic effects could not be transmitted instantaneously, but experiments were inconclusive because of the enormous speed of propagation. The velocity of light is usually described by the letter c.

Maxwell's prediction was tested experimentally by Hertz who generated high frequency alternating currents by means of an induction coil and a spark gap. Electromagnetic waves produced by his apparatus could be reflected and refracted in a manner similar to that of light waves. There was no doubt that electromagnetic energy travelled in waves with the speed of light. This was in 1886; ten years later Marconi was beginning successful experiments on radio propagation. All this arose from Maxwell's reformulation of the circuital law for magnetomotive force. It is one of the most remarkable facts of nature that the velocity of electromagnetic waves in space is the same for waves of all frequencies from low-frequency alternating fields to heat and light waves and beyond to X-rays and γ-rays: a frequency range of more than 1 to 10^{20}.

The solutions of the wave equation can be in terms of any function whatever, i.e. the waves can have any "shape". The most usual function is one that gives a sinusoidal variation of the form $\sin \omega t$, where $\omega = 2\pi f$ and f is the frequency. Since the solution must be of the form $z \pm ct$, we must put $\sin(\omega/c)(z \pm ct)$.

Another way of writing this expression is in terms of wavelength. If the wavelength is λ then the time taken for one wave to pass is λ/c. Hence the frequency is given by

$$f = \frac{c}{\lambda} \tag{8.36}$$

and we can write the expression for the electric field strength as

$$E = E_m \sin \frac{2\pi}{\lambda}(z \pm ct) \tag{8.37}$$

If we are only interested in waves travelling along the positive direction of z

$$E = E_m \sin \frac{2\pi}{\lambda}(z - ct) \tag{8.38}$$

Similarly the magnetic field strength can be written as

$$H = H_m \sin \frac{2\pi}{\lambda}(z - ct) \tag{8.39}$$

The direction of E in our example is along the x axis and the direction of H along the y axis, but the wave moves along the axis of z. Electromagnetic waves are transverse waves and this is well known to be true of light waves. However, it is not true for all waves; pressure waves like those transmitting sound are longitudinal, the pressure is along the direction of propagation.

In general, electromagnetic waves are of course more complicated than the simple example that we have chosen. The electric and magnetic field strengths may have components in all three directions. Our example describes a *plane polarized* wave. Radio aerials are designed to produce polarized waves. Thus television receiving aerials are generally arranged parallel to the polarized electric field.

8.4. ORDERS OF MAGNITUDE IN ELECTROMAGNETIC CALCULATIONS

Maxwell's equations are correct at all frequencies and wavelengths, but it is clear that the order of magnitude of the effects will vary greatly.

Let us first examine the complete statement for the m.m.f. (eqn. 8.1). Suppose we are dealing with a conducting material. The conduction current has a density given by Ohm's law

$$J = \frac{E}{\rho} \tag{8.40}$$

The ratio of this to the Maxwell term is

$$\frac{J}{\varepsilon_0 \dfrac{dE}{dt}} = \frac{E}{\rho \varepsilon_0 \dfrac{dE}{dt}} \tag{8.41}$$

The numerical value of this at an angular frequency ω is $1/\rho\varepsilon_0\omega$. Suppose ρ is of the order $10^{-8}\,\Omega\text{m}$, then the ratio will be of the order $10^{19}/\omega$. It is clear that the conduction current in a metal is enormously bigger than the electric flux term and the latter can easily be neglected. This explains why the m.m.f. around a metallic conductor is dominated by the conduction current and why it is permissible to apply the simplified form of the circuital law even to alternating currents. Of course this is only true if the integral is taken around the conductor.

Let us next examine the wavelength in free space at different frequencies. The wavelength is

$$\lambda = \frac{c}{f} \qquad (8.42)$$

where c is approximately 3×10^8 m/s. If the dimensions of the apparatus are comparable with the wavelength, some interesting results follow. For instance the current is no longer uniform around a circuit. If the circuit is a wavelength long, the current will be flowing in opposite directions over parts of it. Even if the dimensions of the circuit are a good deal smaller than a wavelength, there will still be a variation of current. One cannot then talk about the current in a wire, but only about the current at a particular place in a wire. An arbitrary limit to the size of the circuit in which the current is to be the same at every place is that

$$l < \frac{\lambda}{10} \qquad (8.43)$$

Let us now consider some typical values.

At a power frequency of 50 Hz the wavelength is 6×10^6 m = 6,000 km. It needs a circuit of more than 600 km to violate the condition (8.43). So it is clear that except in the very longest transmission lines there will be no measurable effect.

At a radio frequency of 1 MHz the wavelength is 300m. Thus a circuit should not exceed 30m, which is not an onerous condition. The same is true for a frequency of 100 MHz, which is used for frequency-modulated broadcasting. Here the limiting dimension is 30 cm. However, at 10 GHz the limit is 3 mm and such small circuits are not practical.

Our discussion has shown that the assumptions underlying electric network calculations are correct at all but the highest radio frequencies. This explains the great success of circuit techniques. As long as current is flowing in conductors we generally do not need Maxwell's complete theory to get the right answer.

On the other hand we do need this theory to obtain any sort of answer for problems involving electromagnetic radiation through space. Moreover, although one can go a long way without the complete m.m.f. equation, there are many problems even at low frequencies which cannot be understood without it. A typical example is that of Fig. 8.1.

8.5. A SIMPLE WAVE GUIDE

The usual way of transmitting electromagnetic waves is by means of radio aerials. If these are intended for local broadcasting stations, they are designed to send out energy all around them, but if it is intended to establish point to point communication, aerials can be designed to send out a fairly narrow beam of electromagnetic radiation and this saves energy and stops interference with other communication systems.

However, in all aerial systems the radiation tends to spread out in space and this involves a loss of energy. For transmission over short distances it is often desirable to confine the radiation between conducting surfaces or in tubular conductors. Such devices are called wave guides. We can illustrate this idea by discussing the charging process of a parallel plate capacitor.

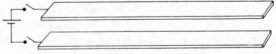

FIG. 8.4 *Charging a large capacitor from one edge*

Figure 8.4 shows a capacitor of large area which is to be charged from a source providing constant p.d. between the plates at one edge. We shall assume that the plates are perfect conductors. Because of *inductance*, i.e. inertia effects, the capacitor cannot be charged instantaneously. It can no longer be regarded as a "capacitance" pure and simple. There will be a charging zone travelling outwards from the source with a finite velocity. Let this velocity be v. Surface charges will appear on the plates and there will be the usual electric field strength between the plates. There will also be a charging current flowing along the plates. Let the

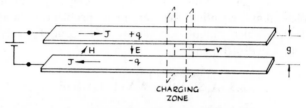

FIG. 8.5 *During the charging process*

charge density be q and the current per unit width be J. Let the magnetic field strength between the plates due to the current be H. The plates of the capacitor can be regarded as a very large loop and application of the circuital law gives $H = J$ A/m. The various quantities are shown in Fig. 8.5. We have

$$J = qv \qquad (8.44)$$

also

$$q = \varepsilon_0 E \qquad (8.45)$$

and

$$B = \mu_0 J \qquad (8.46)$$

whence

$$B = \mu_0 \varepsilon_0 vE \qquad (8.47)$$

Let us apply Faraday's law to a loop embracing the charging zone. To the left of the zone the electric field strength is E and to the right it is zero. The magnetic flux is changing because there is magnetic flux density to the left of the zone and none to the right. Thus

$$\oint E \, dl = -Eg$$

$$= -Bgv \qquad (8.48)$$

so that

$$E = Bv \qquad (8.49)$$

From eqns. (8.47) and (8.49)

$$\mu_0 \, \varepsilon_0 \, v^2 = 1 \tag{8.50}$$

and
$$v = \frac{1}{\sqrt{(\mu_0 \, \varepsilon_0)}} = c \tag{8.51}$$

the velocity of electromagnetic waves. We have shown that the charging process is carried out by an electromagnetic wave travelling between the plates. Instead of looking at these as capacitor plates we can regard them as a wave guide, and we must do so if the dimensions of the plates are comparable with the electromagnetic wavelength.

The observant reader may be surprised that in this example we have discussed electromagnetic radiation without reference to Maxwell's m.m.f. relation. Let us first check whether the result is consistent with Maxwell's theory by taking $\oint H \, dl$ around the charging zone. We have

$$\oint H \, dl = Hl$$

$$= \frac{d\Psi}{dt}$$

$$= l\varepsilon_0 \, Ev \tag{8.52}$$

whence
$$B = \mu_0 \, \varepsilon_0 \, vE \tag{8.53}$$

This is the same as eqn. (8.47) and combined with Faraday's law again results in eqn. (8.51). The results are therefore consistent. The important conclusion follows that Maxwell's discovery of electromagnetic radiation is not an independent statement but is contained in a theory which makes use of forces between charges (ε_0), forces between magnets (μ_0) and Faraday's law.

We have assumed that there are no resistance losses in the walls of the wave guide. However, there is energy being radiated along the guide and this energy must be supplied by the source. Seen from the source the wave guide appears as a resistance

$$Z_0 = \frac{Eg}{Jl} \qquad (8.54)$$

whence $\qquad Z_0 = \frac{Bvg}{lB/\mu_0} = \frac{g}{l}\sqrt{\frac{\mu_0}{\varepsilon_0}}$ ohms $\qquad (8.55)$

This resistance is known as the *radiation resistance* to distinguish it from the dissipative ohmic resistance. The dimensions of $\sqrt{(\mu_0/\varepsilon_0)}$ are also those of resistance and the expression is sometimes referred to as the intrinsic resistance of free space. Its numerical value is 120π, if we take $c = 3 \times 10^8$ m/s.

8.6. THE INTERDEPENDENCE OF ELECTRIC AND MAGNETIC EFFECTS

The subjects of electricity and magnetism grew up more or less independently until Ampère, Faraday and Maxwell showed the close relationship between the two types of phenomena. It is only when we are dealing with static electric or magnetic fields that it is possible to separate the two effects: any problem that deals with changing fields, such as occur with alternating currents or with transients, is neither an electric problem nor a magnetic one, but is electromagnetic in origin. Thus the behaviour of the simple capacitor in the last section during its charging process is governed by inductance as well as capacitance. Many people when asked about the inductance would deny that a capacitor had any such property. Similarly they would be surprised to find that an inductance coil had capacitance. When this capacitance becomes important at high frequencies it is called *stray* capacitance, a grudging admission that there is capacitance which apparently is of an unusual and undesirable kind. It sounds as if there is justifiable annoyance, because one has bought a thing called inductance and finds that this inductance is adulterated with stray capacitance.

The reader of this book will not harbour any such illusions. He will remember that capacitance is associated with the potential energy of electric charges and inductance with their kinetic energy. He will expect interchange between these two forms of energy and will not be surprised that both forms are often associated with the same piece of apparatus. In the early days of radio theory there used to be much discussion about the e.m.f. induced in a coil traversed by a radio wave. This e.m.f. could be worked out by calculating the electric field strength; it could also be derived from the changing magnetic field. The question was asked whether these effects had to be added.

This confusion arose because the electric effect was set over against the magnetic effect, as though there were two waves, one electric and the other magnetic. Our effort to refer both the effects to the mechanical behaviour of electric charges helps us to avoid the confusion. Electricity and magnetism are interdependent. There are neither electric nor magnetic, but only electromagnetic waves.

SUMMARY

The incompleteness of the magnetic circuital law leads to Maxwell's hypothesis that a changing electric flux induces a magnetomotive force. This leads to the discovery that electromagnetic effects travel in transverse waves with the speed of light and that light consists of electromagnetic waves. Energy is transmitted by these waves. Maxwell's equations show that electric and magnetic effects are aspects of the one phenomenon of electromagnetism.

Term	Symbol	Unit (Abbreviation)	Definition
Velocity of electromagnetic waves in free space	c	metres/second (m/s)	$c = \dfrac{1}{\sqrt{(\mu_0 \varepsilon_0)}}$
Wavelength	λ	metres (m)	$\lambda = c/f$
Radiation resistance	Z_0	ohms (Ω)	See Section 8.5

Exercises

8.1. A magnetic flux of uniform density B emerges from the circular pole face of a magnet. The flux density is increasing at a rate B'. What is the acceleration of an electron (charge $-e$, mass m) at rest at a distance r from the centre of the pole face, if no other charges are present? (*Ans.* $B're/2m$)

8.2. A charged capacitor consisting of two concentric spheres discharges symmetrically across the air gap. Determine the magnetic field, using the circuital law with and without Maxwell's term.

8.3. Sea water has a resistivity of about $\rho = 0\cdot36\,\Omega$m and a permittivity $\varepsilon_r = 80$. Find the ratios of the amplitudes of the conduction current density, the polarization current density and the Maxwell term $\varepsilon_0 \, dE/dt$, if E is alternating at 1000 MHz. (*Ans.* $1:1\cdot58:0\cdot02$)

8.4. The capacitor of Fig. 8.5 is fully charged. Discuss in detail what happens if the battery is replaced by a short-circuiting link.

8.5. The capacitor of Fig. 8.5 is fully charged. Discuss in detail what happens if the battery is removed and short-circuiting links are applied to both sides of the capacitor. Show that the behaviour of the circuit is like that of an inductance in parallel with a capacitance.

8.6. A steady current is flowing in a short-circuited inductance of zero resistance. The inductance takes the form of a long flat tube of rectangular cross-section of width a and depth b, where $a \gg b$. The current is J per unit length of tube and flows around the cross-section of the tube.

What happens if the tube suddenly splits slightly all the way across one of its narrow faces? What p.d. appears across the gap? What is the shortest time for the current flow to cease and what is the state of the system at that instant?

(*Ans.* $bc\mu_0 J$, a/c)

8.7. A polarized plane electromagnetic wave travels in a positive direction parallel to the z axis of co-ordinates. The electric field strength is

$$E_x = 1 \cdot 2 \sin \frac{\omega}{c}(z - ct)$$

Find an expression for the magnetic field strength.

$$\left(Ans.\ H_y = \frac{1}{100\pi} \sin \frac{\omega}{c}(z - ct)\ \text{A/m} \right)$$

8.8. The wave in Ex. 7 is being generated by a current flowing in a thin plate which lies in the xy plane at $z = 0$. Another wave of equal amplitude is emitted in the negative z direction. Find an expression for this current.

$$\left(Ans.\ J_x = \frac{1}{50\pi} \sin \omega t\ \text{A/m} \right)$$

8.9. Find the radiation resistance of the thin plate of Ex. 8 per unit length and unit width.

(*Ans.* $60\pi \Omega$)

8.10. Use Maxwell's m.m.f. relationship to find the magnetic field of a slowly moving electric charge and compare it with the magnetic field of a current element. Why is there no need in this proof of Heaviside's conducting fluid? Why is the expression not correct if the velocity of the charge approaches the velocity c?

8.11. Discuss the charging process of Section 8.5, if the space between the parallel plates is filled with a material of permittivity ε_r and permeability μ_r. Derive an expression for the radiation resistance and also for the velocity of the wave front.

$$\left(Ans.\ \sqrt{\left(\frac{\mu_0 \mu_r}{\varepsilon_0 \varepsilon_r}\right)} \frac{g}{l},\ \frac{1}{\sqrt{(\mu_0 \mu_r \varepsilon_0 \varepsilon_r)}} \right)$$

Appendices

APPENDIX A

A LIST OF ELECTRIC AND MAGNETIC QUANTITIES

	Electric	Magnetic
Charge (Pole)	Q	$\bar{Q}$
Dimensional Constant	ε_0	μ_0
Field Strength	E	H
Flux Density	D	B
Susceptibility	χ_e	χ_m
Permittivity (Permeability)	ε_r	μ_r
Potential Difference	V	$\bar{V}$
E.m.f. (M.m.f.)	$\int E\,\mathrm{d}l$	$\int H\,\mathrm{d}l$
Flux	Ψ	Φ
Resistance (Reluctance)	R	$\bar{R}$
Current	I	—
Resistivity	ρ	—
Conductivity	σ	—
Capacitance	C	—
Inductance	—	L, M

A LIST OF USEFUL FORMULAE

Electric	Magnetic	Electromagnetic
$F = \dfrac{Q_1 Q_2}{4\pi\varepsilon_0\, r^2}$	$F = \dfrac{\bar{Q}_1 \bar{Q}_2}{4\pi\mu_0\, r^2}$	$\delta H = \dfrac{I\,\delta l}{4\pi r^2}$
$E = \dfrac{Q}{4\pi\varepsilon_0\, r^2}$	$H = \dfrac{\bar{Q}}{4\pi\mu_0\, r^2}$	$F = IlB_\perp$
$D = \varepsilon_r\varepsilon_0\, E$	$B = \mu_r\mu_0\, H$	$F = QvB_\perp$
$\Psi = \displaystyle\iint D_n\,\mathrm{d}S$	$\Phi = \displaystyle\iint_S B_n\,\mathrm{d}S$	$F_x = NI\dfrac{\partial\Phi}{\partial x}$
$R = \dfrac{\rho l}{S}$	$\bar{R} = \dfrac{l}{\mu_r\mu_0\, S}$	$T_\theta = NI\dfrac{\partial\Phi}{\partial\theta}$
$C = \dfrac{Q}{V}$	$L = \dfrac{\Phi}{I}$	$\displaystyle\oint E\,\mathrm{d}l = -N\dfrac{\mathrm{d}\Phi}{\mathrm{d}t}$
$f_t = \tfrac{1}{2}ED$	$f_t = \tfrac{1}{2}HB$	$\displaystyle\oint H\,\mathrm{d}l = I+\dfrac{\mathrm{d}\Psi}{\mathrm{d}t}$
	$P_h = kfB^x$	
	$W_h = Sl\displaystyle\oint H\,\mathrm{d}B$	$\dfrac{\mathrm{d}^2 E}{\mathrm{d}z^2} = \dfrac{1}{c^2}\dfrac{\mathrm{d}^2 E}{\mathrm{d}t^2}$
	$P_e = \dfrac{2\pi^2 f^2 B_m^2}{3\rho}\, a^2$	$\dfrac{\mathrm{d}^2 H}{\mathrm{d}z^2} = \dfrac{1}{c^2}\dfrac{\mathrm{d}^2 H}{\mathrm{d}t^2}$
		$E = E_m \sin\dfrac{\omega}{c}(z-ct)$
		$\lambda f = c$
		$c = \dfrac{1}{\sqrt{(\mu_0\,\varepsilon_0)}}$

SI AND CGS UNITS

(The velocity c has been taken as 3×10^8 m/s)

	SI	*CGS(e.m.u.)*	*CGS(e.s.u.)*
Force	1 newton	10^5 dynes	10^5 dynes
Energy	1 joule	10^7 ergs	10^7 ergs
Power	1 watt	10^7 ergs/ second	10^7 ergs/ second
Charge	1 coulomb	10^{-1}	3×10^9
Current	1 ampere	10^{-1}	3×10^9
Electric Field Strength	1 volt/metre	10^6	1/30,000
P.D. (E.M.F.)	1 volt	10^8	1/300
Electric Flux Density	1 coulomb/ metre^2	$4\pi \times 10^{-5}$	$12\pi \times 10^5$
Electric Flux	1 coulomb	$4\pi \times 10^{-1}$	$12\pi \times 10^9$
Capacitance	1 farad	10^{-9}	9×10^{11}
Resistance	1 ohm	10^9	$1/(9 \times 10^{11})$
Magnetic Field Strength	1 ampere/ metre	$4\pi \times 10^{-3}$ oersted	$12\pi \times 10^7$
M.M.F.	1 ampere (turn)	4×10^{-1}	12×10^9
Magnetic Pole	1 weber	$10^8/4\pi$	$1/1200\pi$
Magnetic Flux Density	1 tesla	10^4 gauss	$1/(3 \times 10^6)$
Magnetic Flux	1 weber	10^8	1/300
Inductance	1 henry	10^9	$1/(9 \times 10^{11})$
Reluctance	1 ampere/ weber	$4\pi \times 10^{-9}$	$36\pi \times 10^{11}$

SOME USEFUL CONSTANTS

$c = 2\cdot998 \times 10^8$	m/s
$\mu_0 = 4\pi \times 10^{-7}$	H/m
$\varepsilon_0 = 8\cdot854 \times 10^{-12}$	F/m
Electron Charge $= 1\cdot602 \times 10^{-19}$	C
Mass of electron $= 9\cdot108 \times 10^{-31}$	kg
Electron charge/mass $= 1\cdot759 \times 10^{11}$	C/kg
Mass of proton $= 1\cdot672 \times 10^{-27}$	kg

Index

QUEEN MARY
COLLEGE
LIBRARY